Contents

1 Introduction to Semiconductors

Semiconductors are materials whose electrical conductivity lies between that of conductors and insulators. This intermediate conductivity is what makes them so useful in electronic devices. Their behavior is fundamentally tied to their atomic structure and the way electrons are arranged within their crystal lattices. Understanding this behavior is crucial to grasping the function of modern electronics. Think of it as a bridge between the completely free electron flow in metals and the almost complete absence of it in insulators; semiconductors offer a finely tuned control mechanism. This control is precisely what makes them the foundation of the digital revolution.

The crystalline structure of a semiconductor, like silicon or germanium, is a highly ordered arrangement of atoms. These atoms are bonded together in a regular, repeating pattern, forming a three-dimensional lattice. This specific organization is not merely an aesthetic detail but fundamentally dictates how electrons behave within the material. Different crystal structures can lead to different electrical properties, highlighting the close connection between material science and device performance. This lattice structure allows for precise control and manipulation of electron behavior, enabling the creation of sophisticated electronic devices. Visualize this perfect arrangement: it's the stage upon which the electronic symphony is played.

Within this crystalline structure, electrons occupy specific energy levels. Unlike in conductors, where electrons freely move through energy bands, semiconductors possess a significant energy gap—the band gap—between the valence band (where electrons are typically bound to atoms) and the conduction band (where electrons are free to move and conduct electricity). The size of this band gap determines many critical electrical characteristics. A small band gap allows electrons to more easily jump to the conduction band, leading to higher conductivity, while a large band gap makes it more difficult. It is this fine-tuning of the band gap that allows for the diversity of semiconductor applications. The energy landscape governs the electrical behavior; it's the composer's score, dictating the flow of the electronic music.

The ability to manipulate the conductivity of semiconductors is key to their utility. This control is most often achieved through doping, the intentional introduction of impurity atoms into the crystal lattice. Doping with atoms possessing an extra electron (n-type doping) creates an excess of electrons in the conduction band, increasing conductivity. Conversely, doping with atoms with fewer electrons (p-type doping) creates "holes"—the absence of an electron—which act as positive charge carriers. The type and concentration of dopant atoms precisely determine the semiconductor's electrical properties. Imagine carefully selecting instruments in an orchestra - each dopant changes the "sound" of the semiconductor. These charge carriers, electrons, and holes, move under the influence of an electric field. Their mobility—how easily they move through the lattice—is another critical factor influencing conductivity. Several factors impact mobility, including temperature, the type of dopant and its concentration, and the presence of crystal defects. Higher mobility leads to higher conductivity. Temperature plays a significant role; increasing temperature excites more electrons to the conduction band, increasing conductivity in intrinsic semiconductors—those without intentional doping. Understanding the relationship between these factors allows engineers to design semiconductors with specific conductivity properties suited for a range of applications. It's about orchestrating the movement of the charge carriers, controlling their speed and direction for optimal performance.

The interaction between electrons and holes, particularly at the interface between n-type and p-type doped regions—the PN junction—forms the basis of countless semiconductor devices. This junction's behavior is the cornerstone of diodes, transistors, and integrated circuits. This junction isn't merely a physical boundary; it's a dynamic region where charge carriers interact, creating a depletion region where few free charge carriers exist. Applying a voltage across this junction controls the flow of current, allowing for rectification (only allowing current flow in one direction) and amplification. The interplay of these charge carriers at the junction is akin to a complex musical phrase, where each note contributes to the overall harmony.

Therefore, the fundamental principles of semiconductor physics govern the design and operation of all modern electronic devices. From the simplest diodes to the most complex integrated circuits, the interaction of electrons and holes within the semiconductor lattice lies at the heart of it all. This understanding forms the basis for further exploration of more complex semiconductor devices like transistors and integrated circuits, the building blocks of modern electronics. The study of semiconductors is a journey into the heart of the digital age, revealing the elegant physics that drives the technologies shaping our world. Mastering these fundamental concepts is the first step in understanding the intricate world of semiconductor devices.

1.1 Crystalline Structure

The regular arrangement of atoms within a semiconductor crystal is fundamental to its electronic properties. Imagine a perfectly ordered three-dimensional lattice, a repeating pattern extending in all directions. This structure, often described by its lattice constant—the distance between identical lattice points—dictates how electrons move and interact within the material. The precise arrangement varies depending on the semiconductor; silicon, for example, has a diamond cubic structure, where each silicon atom is bonded to four neighboring atoms in a tetrahedral configuration, creating a strong and stable framework. This highly ordered structure is crucial because imperfections or irregularities can significantly impact the semiconductor's performance. Understanding this underlying order is the first step in grasping the behavior of electrons and holes, the charge carriers responsible for semiconductor functionality.

Different semiconductor materials crystallize into distinct lattice structures, impacting their electrical and optical characteristics. Germanium, another popular semiconductor, also exhibits a diamond cubic structure, similar to silicon. However, the subtle differences in their lattice constants and atomic arrangements lead to different bandgaps, influencing their electrical conductivity and suitability for various applications. Variations in the crystalline structure can arise from imperfections such as dislocations, stacking faults, and point defects. These imperfections, often unintended by-products of the manufacturing process, can act as scattering centers, hindering the movement of charge carriers and reducing the overall performance of the semiconductor device. Controlling and minimizing these imperfections is critical for producing high-quality semiconductors.

Crystalline defects are not always detrimental; sometimes, they can be engineered to enhance device performance. For instance, controlled introduction of dopant atoms - impurity atoms intentionally added to the crystal lattice - alters the electrical conductivity of the semiconductor by modifying the concentration of charge carriers. These impurities replace some of the silicon atoms in the lattice, creating either excess electrons (n-type doping) or holes (p-type doping). This controlled alteration of the crystalline structure is the foundation of many semiconductor devices, enabling functionalities like transistors and diodes. The precision required in this process underscores the importance of a thorough understanding of the crystalline structure and its relationship to the material's properties.

The impact of crystallographic orientation on device performance is significant, particularly in the fabrication of integrated circuits. The arrangement of atoms in relation to the surface of the semiconductor wafer significantly influences the growth of subsequent layers during the fabrication process. Misalignment can lead to defects and inconsistencies in the final device. Techniques like epitaxial growth are used to carefully deposit layers of semiconductor material, aligning the crystal structure to the underlying layer, ensuring the integrity of the final structure. This precise control over crystal orientation is essential for producing reliable and high-performance semiconductor devices, highlighting the link between material structure and device functionality.

Furthermore, investigating crystallographic defects through advanced techniques like X-ray diffraction and transmission electron microscopy provides valuable insights into material quality and device reliability. These methods allow researchers to visualize the atomic arrangements, identifying imperfections and their influence on charge carrier transport. By understanding the microscopic features of the crystal structure, researchers and engineers can optimize the fabrication process, improving semiconductor quality and ultimately device performance. This iterative process of analysis and refinement underscores the importance of continued research in crystalline structure and defect engineering within the semiconductor industry. Therefore, the careful study and manipulation of the crystalline structure are indispensable for the development and optimization of semiconductors and related devices. The detailed understanding of crystalline structure lays the crucial groundwork for advanced studies on energy bands and charge carrier behavior.

1.2 Energy Bands

Electrons in an isolated atom occupy discrete energy levels. However, when atoms come together to form a solid, their interactions significantly alter this picture. The outermost electrons, those involved in bonding, are no longer confined to individual atoms; instead, their wavefunctions overlap, leading to a broadening of the atomic energy levels into energy bands. This fundamental concept is crucial for understanding the electrical properties of solids, especially semiconductors.

Imagine a crystal lattice, a highly ordered arrangement of atoms. As these atoms get closer, their outermost electron orbitals begin to interact, overlapping and forming continuous energy bands. These bands, separated by energy gaps, are not continuous spaces but rather regions of allowed electron energies. Electrons can only exist within these bands, a direct consequence of the wave-like nature of electrons and the periodic potential of the lattice.

The valence band is the highest energy band that is completely filled with electrons at absolute zero temperature. These electrons are tightly bound to the atoms and contribute to the material's chemical bonding. Above the valence band lies the conduction band, a band that is usually empty at absolute zero. The energy gap, known as the band gap, separates the valence and conduction bands. The size of this band gap dictates whether a material is a conductor, insulator, or semiconductor.

In conductors, the valence and conduction bands overlap. This means that electrons can easily move from the valence band to the conduction band, even at low temperatures, leading to high electrical conductivity. In insulators, the band gap is very large, typically several electron volts. This makes it extremely difficult for electrons to jump to the conduction band, resulting in very low conductivity. Semiconductors fall somewhere in between. They possess a relatively small band gap, typically on the order of 1 electron volt, allowing electrons to transition to the conduction band with a moderate amount of energy. This transition can be induced by thermal excitation, light absorption, or the addition of impurities.

The concept of effective mass is also crucial in understanding the behavior of electrons within the energy bands. Because of the periodic potential of the crystal lattice, electrons do not behave exactly as free particles. Their motion is influenced by the lattice, leading to an apparent mass, which can be different from their rest mass. This effective mass can be positive or negative, depending on the curvature of the energy bands. A positive effective mass indicates that the electron behaves like a normal particle, while a negative effective mass suggests unconventional behavior. Understanding effective mass is vital for calculating the electron's mobility and conductivity within the material.

Furthermore, the energy band structure can be significantly modified by introducing impurities, a process known as doping. Adding impurities with extra electrons (n-type doping) increases the number of electrons in the conduction band, thereby increasing conductivity. Conversely, adding impurities with fewer electrons (p-type doping) creates "holes" in the valence band, positively charged vacancies that act as mobile charge carriers, also enhancing conductivity. The type and concentration of dopants precisely control the electrical characteristics of the semiconductor. This ability to tailor the properties of semiconductors through doping is essential for creating various semiconductor devices.

The energy band structure provides the framework for understanding the behavior of electrons and holes in semiconductors. It lays the foundation for grasping various phenomena, such as carrier transport, optical absorption, and the operation of semiconductor devices. A thorough understanding of energy bands is, therefore, fundamental to comprehending the workings of the many electronic devices that shape our modern world. The manipulation of energy bands and carrier concentrations is the cornerstone of modern semiconductor technology. By carefully controlling the band gap and doping levels, engineers can design and fabricate a vast array of semiconductor devices with diverse functions, from simple diodes to complex integrated circuits. The beauty lies in the precise control over these subtle quantum mechanical effects, leading to the creation of sophisticated technology.

2 Charge Carriers

Electrons, the negatively charged fundamental particles, and holes, the absence of electrons acting as positive charges, are the heart of semiconductor behavior. Understanding their roles is paramount. In silicon, for instance, at absolute zero, the valence band is completely filled, and the conduction band is entirely empty. This makes silicon an insulator at such low temperatures. However, as temperature rises, thermal energy excites some electrons from the valence band to the conduction band, leaving behind holes. These mobile electrons and holes become our charge carriers, enabling electrical current flow. The number of these carriers directly impacts the semiconductor's conductivity.

Doping, the intentional introduction of impurities into a pure semiconductor crystal, dramatically alters the carrier concentration. Adding a small amount of phosphorus, a pentavalent impurity, introduces extra electrons into the crystal lattice, increasing the number of free electrons—this is n-type doping. Conversely, adding boron, a trivalent impurity, creates 'holes'—empty spaces where electrons should be—leading to p-type doping, where holes are the majority carriers. The precise doping level, meticulously controlled during manufacturing, dictates the electrical characteristics of the resulting semiconductor material, making it possible to tailor materials to specific circuit requirements. Think of it as composing a symphony - each instrument (dopant) adds its unique timbre to the overall sound.

Now let's consider mobility, a measure of how easily charge carriers move through the material under the influence of an electric field. Several factors influence mobility; among the most significant are temperature and the presence of impurities or defects within the crystal structure. Higher temperatures increase thermal vibrations of the lattice atoms, thus scattering charge carriers and reducing their mobility. Similarly, impurities and defects act as scattering centers, impeding the smooth flow of charge. A higher mobility means a greater ability to conduct current, offering design advantages in faster switching speeds and lower power consumption. Consider this: Imagine a highway with no traffic - high mobility. Now imagine rush hour - low mobility. It is a direct analogy to the movement of charge carriers in a semiconductor.

Conductivity, a material property relating the current density to the applied electric field, is directly related to both carrier concentration and mobility. It's the overall measure of how well a material allows electrical current to flow. The formula, usually expressed as = nqn + pqp, neatly sums up this relationship. Here, n and p represent electron and hole concentrations, q is the elementary charge, and n and p are electron and hole mobilities, respectively. Therefore, increasing either the concentration or mobility of the charge carriers increases the semiconductor's conductivity. Engineers use this relationship extensively to design devices with precise conductivity levels for different applications. The formula itself is a concise description of the electrical behaviour of a material, but it also represents the cumulative effect of the underlying microscopic physics at play.

Understanding charge carrier concentration, mobility, and conductivity is fundamental to semiconductor device design. Each parameter significantly affects a device's performance, making a deep understanding of their interplay crucial for building efficient and reliable circuits. The precise control over doping levels provides engineers with a remarkable degree of freedom in tuning the electronic properties of semiconductor materials. This control allows for the creation of sophisticated circuits and devices, performing diverse functionalities that underpin modern electronics. Remember that these three properties work together - you can't consider one without thinking of its interactions with the other two - just like the different sections of an orchestra must work in harmony to produce a beautiful melody. It is this harmony that allows semiconductor technology to create such a wide range of possibilities.

2.1 Electrons and Holes

Electrons, the fundamental negatively charged particles, are key players in the semiconductor story. They reside within the atomic structure, specifically in energy levels and bands as described in the preceding chapters. In a semiconductor, these electrons are situated within the valence band, tightly bound to their respective atoms, happily participating in chemical bonding. However, the game changes when sufficient energy—thermal energy or light, for instance—is supplied. This energy boost excites some electrons, liberating them from their atomic bonds. They then leap into the conduction band, becoming mobile and available to conduct electricity. These liberated electrons represent a significant change in the material's behavior, transforming its electrical properties. Their movement constitutes an electric current, the foundation of many semiconductor devices. Their journey isn't a simple linear progression; it's a complex dance influenced by the material's crystal structure, temperature, and the presence of impurities. Understanding the dynamics of these excited electrons is crucial to grasping the essence of semiconductor functionality. It's essential to note that the number of electrons freed from their bonds is directly related to the energy supplied; more energy equals more free electrons.

Now, imagine the opposite: the absence of an electron where one should be. This absence, a void in the electron sea, acts as a positively charged entity—a "hole." This hole isn't an actual particle, but rather a convenient conceptualization that greatly simplifies analysis of current flow in semiconductors. Think of it as a bubble in a liquid—the absence of liquid in a defined area. A hole represents a lack of negative charge, effectively acting like a positive charge. It can "move" through the crystal lattice, as electrons from neighboring atoms fill the hole. This movement of electrons to fill holes is effectively the movement of the positive charge, meaning the hole moves in the opposite direction of the electrons. This fascinating interplay between electrons and holes dictates how a semiconductor behaves under various conditions, including the application of external electric fields, as we'll explore in later chapters. This concept is crucial for understanding the behavior of semiconductors under various circumstances, making it a central concept in semiconductor physics. It's not just an abstract idea; it is at the heart of many practical applications.

The behavior of both electrons and holes—their movement and interactions within the semiconductor lattice—is influenced significantly by the presence of impurities. These impurities, intentionally introduced through a process called doping, are strategically chosen to enhance the availability of either electrons (n-type doping) or holes (p-type doping). N-type doping, for example, introduces impurities with extra valence electrons, increasing the concentration of free electrons. These extra electrons are not bound to a particular atom, are easily excited into the conduction band and enhance the material's conductivity. Conversely, p-type doping uses impurities that have fewer valence electrons than the host semiconductor. These impurities create "holes," increasing the concentration of positive charge carriers. This manipulation of carrier concentration via doping is a cornerstone of semiconductor device fabrication. The controlled introduction of these impurities permits us to fine-tune the material's properties, adapting it to a wide range of applications, as explained in the following chapters that explain the creation of semiconductor devices. The precise control of doping levels allows engineers to build sophisticated circuits with highly specialized functions.

The concentration of both electrons and holes—what we refer to as carrier concentration—is not static. It's dynamic, intricately linked to temperature. Higher temperatures provide more thermal energy, exciting more electrons into the conduction band, thus increasing both electron and hole concentrations. This change directly impacts the material's electrical conductivity. Similarly, the illumination of a semiconductor with light can also boost carrier concentration, as photons provide energy that liberates electrons from their bonds. This effect is the underlying principle of many optoelectronic devices. Consider the photovoltaic effect, where light generates electricity in a solar cell. It's a direct consequence of light-induced increases in electron and hole concentration. The relationship between carrier concentration and temperature or light exposure is not just an academic curiosity; it has vital practical implications in designing and utilizing semiconductor devices effectively. Therefore, thorough understanding of this relationship is essential for mastering semiconductor physics.

Lastly, we must consider the concept of mobility. Mobility quantifies how easily electrons and holes move through the semiconductor lattice. It's influenced by factors such as temperature, the semiconductor material itself, and the concentration of impurities. A higher mobility indicates that carriers can move more freely, leading to increased conductivity. Similarly, the effective mass of the charge carriers also influences their mobility. Electrons, for instance, in silicon, have an effective mass slightly less than the mass of a free electron. This means electrons in silicon will respond to electric fields more rapidly than free electrons would, which has implications for their mobility. Understanding mobility is crucial for predicting and optimizing the performance of semiconductor devices, such as transistors, which rely heavily on the efficient movement of charge carriers. Therefore, grasping the concepts of electron and hole movement is fundamental to the understanding of semiconductor behaviour and is a necessary tool in the design and analysis of devices.

2.2 Carrier Concentration

Understanding carrier concentration is crucial for mastering semiconductor behavior. It dictates the material's electrical properties, directly influencing its ability to conduct current and its response to external stimuli. Think of it as the volume control for a semiconductor's electrical activity; a higher concentration means a louder signal, a more responsive device. This understanding forms the cornerstone for designing and analyzing semiconductor devices, from simple diodes to complex integrated circuits. The number of charge carriers - electrons and holes - available for conduction isn't a static quantity.

Intrinsic semiconductors, those without intentional impurities, possess a concentration determined by their temperature and the material's bandgap. Increasing temperature excites electrons from the valence band to the conduction band, creating electron-hole pairs. This means a higher temperature generally leads to a higher carrier concentration. The bandgap, on the other hand, represents the energy barrier electrons need to overcome to transition, influencing the number of electrons available for conduction at a given temperature. A smaller bandgap facilitates this transition, leading to higher intrinsic carrier concentration at a specific temperature. The precise relationship between these factors is governed by complex equations, typically involving Boltzmann statistics and exponential functions that account for the probability of electrons occupying energy states. Mastering these relationships allows precise calculation and understanding of intrinsic carrier concentration.

Extrinsic semiconductors, on the other hand, offer a far more controllable carrier concentration. By intentionally introducing impurities, or dopants, we can significantly alter the number of free electrons or holes. This process, called doping, is fundamental to semiconductor technology. Adding pentavalent dopants like phosphorus or arsenic to a silicon crystal introduces extra electrons, creating n-type semiconductors where electrons are the majority carriers. Conversely, trivalent dopants such as boron or gallium create p-type semiconductors with holes as the majority carriers. The concentration of these dopants directly correlates to the carrier concentration; higher doping levels lead to higher carrier concentrations of the majority carrier type.

The level of doping directly impacts the device's electrical properties and its suitability for different applications. Lightly doped semiconductors exhibit higher resistivity, slowing down charge carrier movement and potentially improving their performance in specialized applications. Heavily doped semiconductors, conversely, are characterized by lower resistivity, leading to higher conductivity and enabling their use in high-current applications. Careful consideration of the required doping concentration is crucial in achieving the desired performance in different types of devices. Understanding this directly translates to better device optimization. The choice of the dopant itself also subtly influences other properties which might be beneficial or detrimental depending on the desired performance. The design process needs to account for all these considerations.

Calculating the precise carrier concentration in an extrinsic semiconductor involves considering both the dopant concentration and the intrinsic carrier concentration at the operating temperature. This calculation requires careful consideration of the charge neutrality principle; the total positive charge must equal the total negative charge. This involves complex equations accounting for the contribution of both the majority and minority carriers; this necessitates a good understanding of semiconductor physics principles. Mastering these calculations is fundamental to understanding the device's behavior under various conditions, especially at higher temperatures or under illumination where the generation of electron-hole pairs increases the minority carrier concentration and affects overall device performance.

Beyond simple calculations, understanding carrier concentration also demands an awareness of its impact on other semiconductor properties. Mobility, for example, the ease with which charge carriers move through the material, is affected by carrier concentration. Higher carrier concentrations often lead to decreased mobility due to increased scattering of carriers by each other and by impurities. This interplay between carrier concentration and mobility influences the overall conductivity of the material. Conductivity, a measure of how well a material conducts electricity, is directly proportional to both carrier concentration and mobility. Understanding these interrelationships is essential for comprehending the complete picture of a semiconductor's behavior. A comprehensive understanding of these interactions is vital for accurate device modeling and prediction of its performance.

In conclusion, a deep understanding of carrier concentration is fundamental to mastering semiconductor physics. It's not just a numerical value; it's the key to understanding how semiconductors behave and how they are used to create the electronic devices that underpin modern technology. The implications of controlling and manipulating this variable through doping and temperature control cannot be overstated. From designing simple diodes to building advanced integrated circuits, this chapter's content is essential to the skill set of anyone working with semiconductors.

2.3 Mobility and Conductivity

Electron mobility, a crucial parameter in semiconductor physics, describes how readily electrons move through a material under the influence of an electric field. It's not simply about speed; it's about how effectively they navigate the complex landscape within the crystal lattice. Impurities, crystal defects, and even the temperature of the material significantly impact this mobility, acting as obstacles that scatter the electrons, reducing their overall drift velocity. Understanding these scattering mechanisms is paramount for designing high-performance semiconductor devices. For instance, reducing lattice imperfections through advanced crystal growth techniques directly translates to higher electron mobility and thus improved device performance. The higher the mobility, the faster the electrons can respond to changes in electric fields, resulting in faster switching speeds in transistors and other devices. This has direct implications for the speed and efficiency of modern electronics.

Hole mobility, the counterpart to electron mobility, exhibits similar behavior but with some crucial differences. Unlike electrons, holes are not actual particles but rather the absence of an electron in the valence band. They behave like positively charged particles and contribute significantly to the conductivity of p-type semiconductors. However, their movement is not as straightforward as that of electrons because they essentially represent the movement of the absence of charge. This subtle difference influences scattering mechanisms, affecting the value and temperature dependence of hole mobility. Experimentally determining both electron and hole mobilities is important for accurate device modeling. Different materials and doping levels result in significantly different mobilities, a factor that must be carefully considered during device design and fabrication. The use of advanced characterization techniques helps engineers fine-tune materials and processes for optimum carrier mobility.

Conductivity, a macroscopic property, is directly linked to both electron and hole mobility. It's a measure of how well a material conducts electric current. Higher mobility translates to higher conductivity, meaning the material can carry more current for a given applied voltage. In semiconductors, conductivity is not only dependent on mobility but also strongly influenced by the concentration of charge carriers (both electrons and holes). This concentration is directly controlled by doping, the intentional introduction of impurities into the semiconductor lattice. For example, adding a small amount of phosphorus to silicon increases the number of free electrons, making the material more conductive. Conversely, adding boron introduces "holes," enhancing conductivity in a different manner. The precise balance between these two factors—carrier concentration and mobility—is critical to achieving the desired conductivity in semiconductor devices.

The interplay of temperature and conductivity is a vital consideration. At lower temperatures, scattering due to lattice vibrations is less prevalent, leading to higher mobility and consequently higher conductivity. This temperature dependence is not linear; it's more complex, influenced by the dominant scattering mechanisms at different temperatures. For instance, at very low temperatures, impurity scattering might dominate, while at higher temperatures, phonon scattering becomes the primary factor. This knowledge is crucial in designing devices that operate over a wide range of temperatures, as the performance of a device can significantly degrade due to changes in its conductivity caused by temperature variations. Understanding and managing this temperature dependence is critical in ensuring the reliability and robustness of semiconductor devices operating in diverse environments.

Finally, consider the practical implications of mobility and conductivity. These parameters are not just abstract concepts; they directly translate to the performance characteristics of real-world devices. High mobility semiconductors are essential for high-speed transistors, resulting in faster processors and more efficient electronics. In solar cells, optimizing conductivity is vital for maximizing the current generated from absorbed light. In light-emitting diodes (LEDs), carrier mobility influences efficiency and the overall brightness of the light output. Ultimately, a deep understanding of mobility and conductivity is a cornerstone of modern solid-state electronics, impacting countless technologies that underpin our daily lives. This understanding allows engineers to design devices with improved efficiency, speed, and reliability, shaping the future of electronics.

3 PN Junctions

The PN junction is a fundamental building block in semiconductor electronics. It's formed by joining P-type and N-type semiconductor materials, creating a region of unique electrical properties crucial for many applications. Understanding its behavior is key to mastering the complexities of solid-state devices. Think of it as a tiny, meticulously crafted gatekeeper controlling the flow of electrical current. The magic of the PN junction lies within the depletion region. This is a narrow zone near the junction where mobile charge carriers—electrons in the N-type and holes in the P-type—recombine, leaving behind a region depleted of these carriers. This depletion region acts like a barrier, preventing the free flow of current. The width of this region is crucial and dynamic; it adjusts depending on factors such as applied voltage and temperature. A larger reverse bias voltage expands the depletion region, and vice-versa. The magnitude of the depletion region is a direct result of the interplay between the electric fields generated by the ionized dopant atoms and the concentration of these atoms within the respective regions.

A key characteristic of a PN junction is its rectifying ability. This means it allows current to flow easily in one direction (forward bias) but significantly restricts current flow in the opposite direction (reverse bias). This property is essential for converting alternating current (AC) to direct current (DC), a process known as rectification. Imagine a one-way street for electrons; forward bias opens the gate, and current flows freely. Reverse bias, however, closes it, significantly reducing or blocking the flow. The precise level of current suppression under reverse bias depends on the material properties and the junction's geometry.

This one-way behavior is directly attributable to the potential barrier formed within the depletion region. Under forward bias, the applied voltage reduces this barrier, allowing majority carriers to easily overcome it and contribute to current flow. Conversely, under reverse bias, the barrier increases, substantially limiting current flow to only the small leakage current, primarily due to minority carriers. Therefore, the depletion region's size plays a major role in the junction's electrical characteristics, making it a crucial area for precise control and manipulation. This intricate relationship underscores the importance of understanding the PN junction's physical properties and behavior.

The junction's behavior doesn't just stop at rectification. It also exhibits a phenomenon called breakdown. When a sufficiently large reverse bias voltage is applied, the electric field across the depletion region becomes so strong that it can rip electrons from their atomic bonds, creating electron-hole pairs. This results in a sharp increase in reverse current, potentially damaging the junction. This breakdown voltage, while potentially destructive, is also exploited in various applications. Specific designs, such as Zener diodes, utilize this controlled breakdown to regulate voltage in circuits, offering a stable voltage reference despite current fluctuations. Controlled breakdown has proven to be pivotal in the creation of these precise voltage regulators.

PN junctions have countless applications in electronics. They are fundamental components in diodes, used in rectification, signal clipping, and protection circuits. They are also the foundation for more complex devices like transistors and integrated circuits. The ability to precisely manipulate the material and its characteristics during manufacturing is crucial in achieving the desired behavior and functionality for every application. This control allows for the tailoring of various electrical properties to satisfy a wide range of requirements. From simple rectification tasks in power supplies to intricate signaling and amplification roles within integrated circuits, these ubiquitous devices stand as a testament to our ability to engineer the material world at the atomic level. Their reliability and versatility are essential factors contributing to the ubiquitous nature of these devices within the modern electronics landscape.

The simplicity of the PN junction belies its profound impact on modern technology. From the simple rectifier in your phone charger to the complex integrated circuits in your computer, this foundational element plays a critical role. By understanding its behavior, you gain access to a deep understanding of how many essential components of our modern electronics work. It provides a strong basis for comprehending the more sophisticated semiconductor devices we will explore in subsequent chapters. The mastery of this simple yet powerful building block is essential to unlocking the deeper mysteries of the semiconductor world.

3.1 Depletion Region

The depletion region, a seemingly simple concept, is the heart of many semiconductor devices. It's formed at the junction of P-type and N-type semiconductors, a boundary where the magic of solid-state electronics truly begins. Imagine two distinct territories, one rich in holes (positive charge carriers) and the other teeming with electrons (negative charge carriers). When these regions meet, a fascinating interaction unfolds; electrons from the N-type diffuse into the P-type, while holes from the P-type diffuse into the N-type. This diffusion, driven by the inherent desire for equilibrium, isn't a chaotic scramble; it's a carefully orchestrated movement, a delicate dance between opposing forces.

This diffusion process leaves behind uncovered ions. In the N-type region, negatively charged acceptor ions are left behind, creating a region depleted of mobile charge carriers, thus giving its name to this zone: the depletion region. Conversely, in the P-type region, positively charged donor ions remain, further contributing to the depletion. This isn't just a simple loss; it's the creation of an electric field, a barrier that opposes further diffusion. This field, internal to the junction, acts like a gatekeeper, meticulously controlling the flow of charge carriers. It's a self-regulating system, perfectly balancing the desire for equilibrium against the electric field's resistance.

The width of this depletion region is dynamic, influenced by factors like the doping concentration of both P and N regions and the applied external voltage. A higher doping concentration leads to a narrower depletion region, while a lower concentration results in a wider one. This sensitivity to doping provides engineers with a powerful tool for fine-tuning the characteristics of semiconductor devices. Similarly, an external reverse bias voltage widens the depletion region, strengthening the internal electric field, creating a stronger barrier to current flow. Conversely, a forward bias voltage narrows the depletion region, reducing the barrier and allowing current to flow. Understanding this relationship is essential for comprehending the behavior of diodes and other semiconductor devices.

The depletion region's influence extends beyond simply regulating current flow. It's the key to understanding the rectifying behavior of a PN junction, the ability to allow current flow in one direction while effectively blocking it in the other. This unidirectional behavior is fundamental to numerous electronic applications, from simple power supplies to sophisticated integrated circuits. Without the depletion region's control, the chaotic movement of charge carriers would render the construction of efficient and controlled electronic circuits impossible. It's the silent conductor, subtly orchestrating the electrical symphony within the device.

Consider the analogy of a dam on a river. The water (charge carriers) wants to flow downstream (from the N-type to the P-type in forward bias), but the dam (the depletion region) initially prevents its flow. When we open a gate (apply a forward bias), the water flows freely, but closing the gate (applying a reverse bias) halts the flow completely. This straightforward analogy demonstrates how the depletion region acts as a voltage-controlled gate, regulating the passage of current. The size of the gate, in the context of a depletion region, directly relates to the applied voltage. Larger gates facilitate the flow, whereas smaller gates restrict it significantly.

The depletion region's impact extends to more complex devices, such as bipolar junction transistors and field-effect transistors. In these devices, the careful control of the depletion region's width and electric field is crucial for their amplification capabilities, creating the foundation of modern electronics. The precise manipulation of the depletion region through doping, external voltages, and physical structure allows engineers to design circuits with remarkable complexity and performance. The intricate dance of electrons and holes within this region governs a vast array of technologies we rely on every day.

In summary, the depletion region is not merely a passive space; it's a dynamic and crucial component. Its properties are directly controllable, making it a powerful tool for engineers. It's the cornerstone of countless electronic devices, highlighting the importance of understanding its behavior to truly grasp the functionality and operation of semiconductors. It is the silent heart of semiconductor technology. A thorough grasp of the depletion region's nuances will enable you to navigate the complexities of solid-state electronics with confidence, transforming seemingly abstract concepts into tangible understanding, thus paving the way for innovative advancements in the field. Mastering the depletion region is crucial for unlocking the potential of semiconductor devices and the future of electronics. Its significance is truly undeniable.

3.2 Diode Characteristics

The diode, a fundamental building block in electronics, owes its unique properties to the PN junction. This junction, formed by joining P-type and N-type semiconductor materials, creates a depletion region—an area devoid of free charge carriers. The depletion region acts as a barrier, preventing the flow of current unless a sufficient external voltage is applied. Understanding the behavior of this re-

gion is key to grasping diode characteristics.
Consider a typical silicon diode. When a forward bias voltage is applied—positive terminal to the P-side, negative to the N-side—the depletion region narrows. This reduction in the barrier allows majority carriers—holes in the P-side and electrons in the N-side—to overcome the barrier and flow across the junction, resulting in a substantial current. The relationship between current and voltage isn't linear; it follows an exponential function, described by the Shockley diode equation: I = Is(e(V/nVT) \item 1), where I is the diode current, Is is the reverse saturation current, V is the applied voltage, n is the ideality factor (typically between 1 and 2), and VT is the thermal voltage.
Conversely, applying a reverse bias—negative terminal to the P-side, positive to the N-side—widens the depletion region. This increased barrier significantly restricts the flow of majority carriers. A small reverse saturation current does exist, however, primarily due to the movement of minority carriers. This reverse current is temperature-dependent, increasing with higher temperatures, a characteristic useful for certain temperature-sensing applications.
The diode's I-V characteristic curve visually represents this behavior. The forward bias region shows a rapid increase in current with a small increase in voltage, while the reverse bias region reveals a very small, essentially constant, reverse saturation current until the breakdown voltage is reached. Exceeding the breakdown voltage leads to a sharp increase in reverse current, potentially damaging the diode if not properly managed. This breakdown can be caused by either avalanche breakdown, where carriers gain enough energy to ionize other atoms, creating an avalanche effect, or Zener breakdown, resulting from a high electric field across the depletion region.
Analyzing the diode's behavior requires considering several factors. Temperature significantly affects both forward and reverse characteristics. Increasing temperature increases the reverse saturation current and lowers the forward voltage drop for a given current. This necessitates careful consideration of operating temperature ranges when designing circuits. Additionally, different diode

types, such as Schottky diodes or Zener diodes, exhibit distinct characteristics, catering to different applications. Schottky diodes, for example, are known for their fast switching speeds, while Zener diodes are specifically designed for voltage regulation.

Furthermore, the diode's dynamic resistance, defined as the change in voltage divided by the change in current, varies significantly over its operating range. In the forward bias region, this resistance is low, while in the reverse bias region, it is extremely high. Understanding the dynamic resistance is crucial for accurate circuit analysis and predictions, particularly when considering the impact of AC signals superimposed on a DC bias. Precise modeling often goes beyond the simplified Shockley equation, employing more sophisticated models that account for additional effects such as series resistance, capacitance, and temperature dependencies.

The practical implications of diode characteristics are vast. Their rectifying ability forms the basis for power supplies, converting alternating current (AC) to direct current (DC). Moreover, they serve as switches, logic gates, and signal limiters, allowing for diverse functionalities within electronic circuits. Careful selection of diode type and operating conditions is crucial for ensuring efficient and reliable performance in any application. Choosing an incorrect diode for a particular application can lead to malfunctions, reduced efficiency, or even component damage. Understanding the fundamentals—forward and reverse bias behaviors, the relationship between current and voltage, and the role of temperature and different diode types—is therefore essential for any practicing engineer or physicist working with semiconductors. The intricacies of the PN junction are far-reaching, impacting many areas of electronic design and function.

3.3 Rectification

The PN junction's unique ability to rectify, or convert alternating current (AC) to direct current (DC), stems directly from the built-in potential difference across the depletion region. This potential barrier, formed by the diffusion of majority carriers across the junction, acts as a one-way valve for current flow. When a forward bias is applied—meaning the positive terminal of the external voltage source is connected to the P-side and the negative to the N-side—the barrier is reduced, allowing a significant current to flow. Electrons from the N-side readily overcome the lowered barrier and recombine with holes on the P-side, leading to a substantial current.

Conversely, under reverse bias, the situation is dramatically different. The external voltage source increases the width of the depletion region, effectively widening the potential barrier. This makes it exceedingly difficult for majority carriers to cross the junction, resulting in a negligible current flow - only a tiny reverse saturation current, predominantly due to minority carrier movement. This asymmetric behavior forms the foundation of rectification. It's a fundamental principle that underpins many essential electronic devices.

Consider a simple diode circuit: you apply an AC voltage source across the diode. During the positive half-cycle of the AC wave, the diode is forward-biased, permitting current flow. During the negative half-cycle, however, it's reverse-biased, blocking current. The resultant output is a pulsating DC signal, a rectified version of the input AC signal. This simple yet powerful concept is why diodes are integral components in power supplies, converting household AC power to the DC needed for electronic devices.

The effectiveness of a diode's rectification is quantified by its rectification ratio, a measure of the difference between forward and reverse current. A higher rectification ratio indicates better rectification. Factors influencing this ratio include the diode material, temperature, and the applied voltage. Higher temperatures generally lead to a slightly lower rectification ratio due to increased thermal generation of minority carriers. The choice of semiconductor material significantly impacts the diode's characteristics, with silicon and germanium being the most common choices, each offering a unique performance profile. Silicon diodes, for example, generally exhibit a higher rectification ratio compared to germanium diodes.

The process of rectification is not merely about converting AC to DC; it's also about controlling the flow of electrical energy. Think of it as a sophisticated gatekeeper, selectively allowing current to pass in only one direction, thereby shaping and controlling the electric signals within a circuit. Without rectification, many of our modern electronic marvels, including your phone charger, computer power supply and numerous other electronic devices, simply wouldn't function. Imagine a world without the ability to effortlessly convert AC to the DC power that fuels our modern lives; it would be a world vastly different from the technologically advanced society we inhabit today.

The efficiency of a rectifier circuit isn't perfect. The non-ideal behavior of real diodes leads to power loss, primarily due to the voltage drop across the diode during forward bias. To mitigate this power loss and improve efficiency, several rectifier configurations have been developed, including half-wave rectifiers, full-wave rectifiers, and bridge rectifiers. Each configuration has its own advantages and disadvantages regarding efficiency, complexity, and output waveform quality. The selection of a particular rectifier circuit depends heavily on the specific application and its requirements for efficiency and waveform characteristics. Understanding these factors is critical to the design of effective power supply systems.

Beyond the simple diode, more complex semiconductor devices, like thyristors and silicon-controlled rectifiers (SCRs), offer more sophisticated control over rectification. These devices allow for controlled switching of the current flow, enabling applications such as power regulation and phase control. The flexibility offered by these advanced components greatly expands the possibilities for manipulating and regulating electric power in diverse systems. These devices aren't just passive rectifiers, they are active components capable of intricate control and shaping of power distribution.

Finally, while the principles of rectification are relatively simple, the practical application and design of rectifier circuits often involves detailed consideration of several factors. The ripple voltage, a residual AC component in the rectified output, needs to be minimized through filtering techniques. This involves the use of capacitors and inductors to smooth the pulsating DC waveform, leading to a cleaner, more stable DC supply. Proper design considerations also account for peak inverse voltage (PIV), the maximum reverse voltage a diode must withstand without failure. Failure to consider PIV can lead to diode breakdown and circuit malfunction. Understanding and addressing these factors is essential for the reliable operation of any rectifier-based circuit.

3.4 Breakdown

The PN junction, a seemingly simple structure, exhibits a fascinating complexity when subjected to sufficiently high reverse bias. This reverse bias, pushing electrons from the p-side towards the n-side and holes in the opposite direction, gradually depletes the region near the junction of its mobile charge carriers. This depletion region, initially thin, widens as the reverse bias increases, creating a strong electric field across it.

However, this widening isn't infinite. As the reverse bias voltage continues to climb, a critical point is reached. At this voltage, the electric field within the depletion region becomes so intense that it can rip electrons from their bonds in the semiconductor lattice, generating electron-hole pairs. These newly created carriers are then swept across the junction by the strong electric field, contributing to a sudden increase in reverse current. This phenomenon is known as avalanche breakdown.

Imagine it like this: the electric field is a powerful wind, and the electrons are leaves clinging to branches. With a gentle breeze (low reverse bias), the leaves stay put. But with a hurricane-force wind (high reverse bias), the wind's strength becomes so immense that it violently tears the leaves from their branches, creating a torrent of leaves being swept away—electrons being swept across the junction.

Another breakdown mechanism, Zener breakdown, manifests itself at lower voltages, primarily in heavily doped junctions. In heavily doped semiconductors, the depletion region is exceedingly narrow. The extremely high electric field across this tiny region can directly tunnel electrons from the valence band of the p-side to the conduction band of the n-side. It's a quantum mechanical effect, where electrons can essentially "jump" through the energy barrier, rather than being ripped free by the field. This tunneling effect leads to a significant increase in reverse current, causing breakdown.

It's crucial to understand that these breakdown mechanisms aren't destructive in themselves, assuming the current is kept within safe limits. Devices are designed to withstand the reverse bias voltage required for avalanche or Zener breakdown—for this reason, these devices aren't merely intended for rectification, and they possess a specific application in the field of electronics. Controlled breakdown is actually exploited in many practical applications. Zener diodes, for instance, leverage Zener breakdown to provide a precise voltage reference in circuits, acting as voltage regulators. Their carefully controlled breakdown voltage ensures a stable output voltage regardless of fluctuations in the input.

The impact of temperature on breakdown voltage is also significant. Generally, the avalanche breakdown voltage decreases with increasing temperature. This is because higher temperatures increase the thermal generation of electron-hole pairs, lowering the electric field needed to initiate avalanche breakdown. Understanding this temperature dependence is essential for reliable circuit design, preventing unexpected behavior at extreme temperatures.

The selection of the right material and the meticulous control of doping concentration during fabrication are key in designing diodes with specific breakdown voltages. Silicon is a common choice, but other materials like germanium and gallium arsenide offer unique properties suitable for specialized applications. A deeper understanding of materials science complements the knowledge of the physics behind PN junction breakdown.

Let's consider a practical example. Imagine designing a power supply that needs a stable 5-volt output. A Zener diode with a 5-volt breakdown voltage could be used to regulate the output, maintaining the voltage even if the input fluctuates. The diode would conduct only when the voltage rises above 5 volts, effectively clamping the output. This is the practical significance of understanding and controlling the breakdown phenomena in PN junctions. The careful consideration of both material selection and the effect of temperature on breakdown voltage is key to ensuring that the circuit behaves precisely as intended. Therefore, the understanding of these mechanisms is essential to fully grasp the capabilities of these devices. Such precise control demonstrates the elegant interplay between physics and engineering that allows us to harness the properties of semiconductors to create useful electronic components. Moreover, without a thorough understanding of these mechanisms, designing robust and reliable electronic systems would be much more difficult.

3.5 Applications

The PN junction, as we've established, is the cornerstone of many semiconductor devices. Its rectifying properties, arising from the built-in potential and depletion region, enable the controlled flow of current, forming the basis for diodes, a vital component across numerous applications. Consider, for instance, the ubiquitous power supply. Here, diodes act as rectifiers, converting the alternating current (AC) from the mains supply into the direct current (DC) needed to power electronic devices. This is a fundamental application, showcasing the diode's crucial role in shaping the electrical landscape around us. Without the precise control over current flow provided by the PN junction, many of our modern conveniences would simply not function. The efficiency of this conversion is directly impacted by the characteristics of the diode, highlighting the importance of understanding the detailed physics of the PN junction to optimize performance and minimize energy loss. Moreover, the design and selection of specific diode types within a power supply—considering factors such as voltage rating, current handling capacity, and switching speed—all depend on a thorough grasp of PN junction behavior.

Beyond simple rectification, PN junctions find sophisticated uses. Zener diodes, specifically designed for operation in the reverse breakdown region, provide voltage regulation, ensuring a stable voltage output despite variations in input voltage or load current. This application is essential for numerous electronic circuits, where maintaining a constant voltage is crucial for the reliable operation of sensitive components. The breakdown voltage of a Zener diode is determined by the doping concentration and the width of the depletion region, demonstrating again the intimate relationship between material properties and device functionality. Precise control over these parameters during fabrication allows engineers to tailor Zener diodes for specific voltage regulation applications. From small-scale consumer electronics to large industrial power supplies, Zener diodes safeguard the delicate balance needed for reliable circuit operation.

Light-emitting diodes (LEDs) represent a visually striking and energy-efficient application of PN junctions. When electrons and holes recombine within the depletion region, energy is released in the form of photons—light. The color of the light emitted depends on the bandgap energy of the semiconductor material used. Therefore, by selecting different semiconductor materials, LEDs can be manufactured to emit light across the entire visible spectrum, and even beyond, into infrared and ultraviolet regions. This versatility makes LEDs suitable for a vast range of uses, from simple indicator lights to advanced display technologies. Furthermore, the energy efficiency of LEDs, far surpassing traditional incandescent bulbs, makes them a vital component in modern lighting systems. The development of high-brightness LEDs has revolutionized lighting technology, highlighting the potential for further innovations based on a deeper understanding of PN junction behavior.

Beyond LEDs, other optoelectronic devices utilize the PN junction. Photodiodes, for example, exhibit the inverse process—they convert light into an electrical signal. The absorption of photons creates electron-hole pairs, generating a current proportional to the light intensity. This principle forms the basis for a wide range of applications, from optical communication systems to light sensors in various industrial and consumer devices. The sensitivity and spectral response of photodiodes are critically linked to the material properties and the design of the PN junction, underscoring the importance of detailed knowledge. From the simple act of detecting the presence of light to complex optical communication systems, photodiodes play a crucial role in many modern technologies.

The versatility of the PN junction extends even further. It lies at the heart of more complex semiconductor devices, such as bipolar junction transistors (BJTs) and thyristors. BJTs, covered in subsequent chapters, represent another significant step in semiconductor technology, building upon the fundamental principles of the PN junction to achieve current amplification. The understanding of PN junctions is fundamental to grasping the workings of the more complex transistor architecture, emphasizing the interconnectivity of concepts within the field of semiconductor physics. Similarly, thyristors, used in power control applications, rely on the intricate interplay of multiple PN junctions to enable high-power switching capabilities. In essence, the PN junction forms the bedrock of a vast landscape of electronic devices, and mastering its behavior is paramount to comprehending the complexity of solid-state electronics.

4 Bipolar Junction Transistors

The bipolar junction transistor (BJT), a cornerstone of modern electronics, stands as a testament to the power of semiconductor physics. Its relatively simple structure belies its remarkable ability to amplify electrical signals, a function crucial to countless applications, from simple amplifiers to complex computer processors. At its heart, the BJT is a three-terminal device comprising two PN junctions back-to-back, forming either an NPN or PNP configuration. This arrangement allows for the control of a larger current flowing between two terminals (collector and emitter) by a much smaller current injected into the third terminal (base). Understanding this intricate interplay is key to grasping the transistor's functionality.

Consider the NPN configuration: A small increase in base current induces a significantly larger increase in collector current. This current amplification, often represented by the common-emitter current gain (), is the core principle behind the BJT's amplifying capability. Think of it as a valve controlling water flow: a small twist of the valve (base current) can significantly alter the water flow (collector current). This proportional relationship, however, isn't perfectly linear across all operating regions; saturation and cutoff regions represent distinct operational behaviors where the relationship deviates from the ideal. Careful consideration of these different operating conditions is vital in circuit design.

Understanding the BJT's behavior necessitates a deeper look into the physics of the PN junctions. The depletion regions at these junctions play a crucial role in controlling current flow. The width of these regions varies depending on the applied voltages, affecting the ease with which charge carriers can traverse the junctions. The base-emitter junction acts as a forward-biased diode, injecting minority carriers into the base region. These minority carriers then diffuse across the base, recombining with majority carriers, and ultimately contributing to the current flowing into the collector. The base-collector junction, conversely, is usually reverse-biased, creating a significant potential barrier. However, the injected minority carriers from the base successfully overcome this barrier, greatly impacting the overall current.

The characteristics of a BJT are typically illustrated using various graphical representations. The common-emitter configuration, one of the most frequently used, shows the relationship between the collector current (Ic), the base current (Ib), and the collector-emitter voltage (Vce). This allows for an easy visualization of the transistor's amplification properties and its operating regions. Families of curves, each corresponding to a specific Ib, help in understanding how the collector current changes with both base current and collector-emitter voltage. This graphical representation proves essential in selecting appropriate operating points for circuit design and analysing transistor behavior in various applications. Detailed analysis of these curves reveals the intricacies of the transistor's operating regions, including active, saturation and cutoff modes.

Furthermore, several key parameters characterize the BJT's performance. Beta (), as previously mentioned, represents the current amplification factor. Alpha () describes the ratio of collector to emitter current. These parameters, alongside input and output resistances, are critical in circuit design to ensure the transistor operates within its intended parameters and achieves desired amplification. It's important to understand that these parameters vary slightly with temperature and operating conditions, necessitating consideration during design. Careful selection of transistors and biasing components is necessary for stable operation.

Designing circuits incorporating BJTs requires understanding biasing techniques. Biasing establishes the proper operating point, setting the DC currents and voltages within the active region of the transistor. Common biasing techniques include voltage divider bias and emitter bias, each offering varying levels of stability and simplicity. Careful choice of biasing method is essential for achieving optimal performance and stability, and depends on the specific requirements of the circuit. Incorrect biasing can lead to unpredictable operation and potentially damage the transistor. Understanding the different configurations, and the pros and cons of each method, is crucial.

Finally, beyond simple amplification, BJTs are fundamental components in various circuit topologies. They are indispensable in switch applications, where they can act as high-speed switches controlling larger currents with relatively small control signals. They are also critical in oscillator circuits, generating periodic signals, and in logic circuits that form the basis of digital systems. The versatility of BJTs stems from their inherent ability to amplify signals, creating an array of applications previously unimaginable without them. As you delve deeper into electronics, the ubiquitous presence of BJTs in even the most complex circuits will become self-evident.

4.1 Structure and Operation

The bipolar junction transistor (BJT), a cornerstone of modern electronics, operates on the principle of current amplification. Understanding its structure is crucial to grasping its function. Imagine two back-to-back diodes—that's essentially what a BJT is, but with a clever twist. It possesses three distinct regions: the emitter, base, and collector, each with unique doping levels. The emitter is heavily doped, providing a copious supply of charge carriers. The base is lightly doped, creating a narrow region, and the collector is moderately doped, forming the current collecting region. The interaction of these regions, their relative doping concentrations and the resulting electric fields are paramount to the device's functionality.

This clever arrangement allows for current control. A small current injected into the base region modulates the flow of a significantly larger current between the emitter and collector. This is the essence of transistor action—a small signal controlling a larger one. Think of it as a valve controlling water flow: a small twist of the valve handle (base current) dramatically changes the amount of water flowing through the pipe (collector current). This is what makes BJTs so valuable in amplification circuits. The physics behind this is the interplay between majority and minority carriers, a concept explored in earlier chapters. It's vital to remember that a BJT is a current-controlled device, in contrast to the voltage-controlled field-effect transistors we'll discuss later.

The operation of a BJT depends on the biasing conditions. In the common-emitter configuration, for instance, the emitter is forward-biased with respect to the base (positive voltage on the base, negative voltage on the emitter) enabling significant current flow from emitter to collector. The collector-base junction however is reverse-biased (positive voltage on collector, negative voltage on base), allowing a collector current significantly larger than the injected base current to flow. It is the subtle interplay between these two junctions and this biasing that sets the stage for amplification. Different bias configurations yield differing operational characteristics, affecting the BJT's behavior and making it suitable for various applications. It is essential to understand these operating regions for effective circuit design.

Now let's consider the various operating regions. There's the active region, where the transistor functions as an amplifier; the saturation region, where it acts like a closed switch; and the cutoff region, where it behaves like an open switch. Understanding these regions is essential to design circuits that function correctly. In the active region, we exploit the transistor's current amplification capabilities. In saturation, the transistor is fully "on," providing a low-resistance path for current flow, and in cutoff, it is completely "off," blocking current flow. These operating modes find use in various circuits, determining whether we're amplifying a signal, switching a circuit on or off, or something in between. Selecting the correct operating region depends on the application. The transistor's performance in each region is characterized by parameters like current gain, which describe the amplification factor for a specific configuration and the operating point of the device.

Beyond simple amplification, BJTs find uses in various more complex circuits. They're crucial components in many analog circuits such as audio amplifiers and oscillators, where their current amplification capabilities are vital. Moreover, they are frequently used in digital logic circuits, serving as switches in various logic gates. It is important to remember that the behavior of a BJT is intrinsically tied to the properties of the semiconductor material used in its construction and the way these are combined. Consequently, variations in these parameters lead to inconsistencies in performance, underlining the importance of careful design and manufacturing processes. In this context, an understanding of the physics is vital. Each application necessitates understanding the relevant operating region and the parameters that govern transistor behavior in that region. Furthermore, understanding the limitations of these transistors in high-frequency operation is critical.

The inherent limitations of BJTs, particularly at high frequencies, stem from the device's physical structure. Transit time effects—the time it takes for charge carriers to traverse the base region—become significant, leading to reduced gain and increased phase shift at higher frequencies. Parasitic capacitances, inherent to the device's structure, also limit high-frequency operation. These capacitances shunt the signal, reducing the transistor's effectiveness. Consequently, BJTs often have limitations in high-frequency applications. Careful selection of transistor types is essential for designing circuits that operate reliably across a wide frequency range. Consideration should be given to advanced techniques like emitter degeneration or feedback that compensate for these limitations and enable use in high-frequency applications. This underlines why, for high-speed applications, Field-Effect Transistors often offer advantages.

4.2 Amplification

The bipolar junction transistor (BJT), a cornerstone of modern electronics, stands as a testament to the power of semiconductor physics. Its ability to amplify weak signals into stronger ones underpins countless applications, from audio amplifiers in our smartphones to complex signal processing in high-speed communication networks. Understanding how this amplification occurs is crucial for anyone seeking a deeper understanding of solid-state electronics.

At the heart of BJT amplification lies the control of current flow. Unlike a simple resistor, the BJT doesn't simply offer passive resistance; it actively regulates the current passing through it. This regulation is achieved through the precise manipulation of charge carriers within its carefully structured layers - the emitter, base, and collector. Imagine a water valve: a small movement of the valve handle (the input signal) controls a much larger flow of water (the output signal). Similarly, a small change in the base current of a BJT can drastically alter the collector current.

This control mechanism hinges on the transistor's unique structure. The three regions - emitter, base, and collector - are meticulously doped to create distinct regions with different charge carrier concentrations. The base region, specifically, is extremely thin and lightly doped compared to the emitter and collector. This design is not arbitrary; it's the key to the amplification effect. The lightly doped base acts as a control gate, regulating the flow of majority carriers (electrons in an NPN transistor, holes in a PNP transistor) injected from the emitter into the collector region. A small change in base current significantly influences the number of carriers successfully traversing this narrow gate.

The amplification process itself can be broken down into a series of steps. First, a small input current is applied to the base, modifying the potential barrier within the base-emitter junction. This alteration in potential affects the flow of majority carriers from the emitter into the base. Crucially, because the base is thin and lightly doped, a significant portion of these carriers will successfully transit the base region without recombining with minority carriers. These carriers are then swept into the collector region by the potential difference applied between the collector and the emitter. Thus, a small base current controls a substantially larger collector current. The ratio between the collector current and the base current is known as the current gain (), a critical parameter that defines the transistor's amplification capability. A high indicates that a small input current can control a significantly larger output current, resulting in substantial signal amplification.

Various circuit configurations exploit this amplification property. The common emitter configuration, for instance, is frequently used because of its high current gain and voltage gain. Here, the input signal is applied to the base, and the output signal is taken from the collector. The emitter is grounded, providing a reference point. Other configurations, like common collector and common base, exist, each with its unique characteristics and applications. The choice of configuration depends largely on the desired amplification characteristics, input and output impedance, and overall circuit design requirements.

Understanding the operating regions of the BJT is critical for effective circuit design. The three main operating regions are the active region, saturation region, and cutoff region. The active region is where the transistor operates as an amplifier, ensuring the desired amplification effect. In contrast, the saturation region represents a state of maximum collector current, which is not always ideal for amplification purposes. The cutoff region, on the other hand, represents the absence of collector current, effectively turning the transistor "off". Careful consideration of these operating regions is crucial for properly designing and optimizing BJT-based amplifiers.

Different BJT types cater to different applications. Signal transistors are optimized for high-frequency applications, prioritizing speed and bandwidth. Power transistors, conversely, are designed to handle large currents and voltages, often found in power amplifier circuits. Selecting the appropriate transistor for a particular application requires careful consideration of its specifications, including its current gain, maximum power dissipation, and frequency response. The seemingly simple transistor, therefore, is a marvel of miniaturized engineering; a testament to the profound impact of our understanding of semiconductor physics. Its ability to amplify signals forms the backbone of countless electronic devices. Mastering this principle is vital for anyone venturing into the realm of solid-state electronics.

5 Field-Effect Transistors

Field-effect transistors (FETs) represent a significant advancement in semiconductor technology, offering distinct advantages over bipolar junction transistors (BJTs). Unlike BJTs, where current flow is controlled by injecting minority carriers, FETs utilize an electric field to modulate the conductivity of a channel, leading to higher input impedance and lower power consumption. This makes them particularly well-suited for various applications, from integrated circuits to high-frequency amplifiers.

Their operation hinges on the principle of controlling a current flow between two terminals, the source and drain, by varying the voltage applied to a third terminal, the gate. This control is achieved through the formation of a depletion region or an inversion layer within the semiconductor material, effectively acting as a gate, controlling the channel's conductance and hence the current.

The most prevalent type of FET is the Metal-Oxide-Semiconductor Field-Effect Transistor (MOSFET). Its structure comprises a semiconductor substrate, a thin insulating oxide layer (typically silicon dioxide), a metallic gate electrode, and source and drain regions diffused into the substrate. Applying a positive voltage to the gate of an n-channel MOSFET repels holes in the substrate, creating an n-type channel allowing current flow between the source and drain. Conversely, in a p-channel MOSFET, a negative gate voltage induces a p-type channel. The simplicity of its construction and its inherent scalability make the MOSFET the workhorse of modern integrated circuits.

In contrast to the MOSFET, the Junction Field-Effect Transistor (JFET) employs a pn junction to control channel conductivity. A JFET consists of a channel of n-type or p-type semiconductor material with two p-type or n-type regions, respectively, forming the gate. Applying a reverse bias to the gate-channel junction depletes the channel, thus reducing its conductivity. The current flowing between the source and drain is consequently modulated. JFETs are generally simpler to manufacture compared to MOSFETs, particularly at smaller scales, though MOSFETs' superior characteristics regarding current control and lower leakage currents have made them dominant in many applications.

The choice between MOSFETs and JFETs depends largely on the specific application. MOSFETs exhibit a higher input impedance, making them ideal for applications requiring minimal loading on preceding stages of a circuit. Their superior characteristics also lead to less signal distortion and enable higher circuit density in integrated circuits. On the other hand, JFETs boast a lower noise figure, making them preferable for certain high-frequency and low-noise applications. This advantage stems from the absence of a gate oxide layer and associated interface states, which are a source of noise in MOSFETs. Furthermore, JFETs generally have simpler manufacturing processes, particularly for older technologies, making them potentially more cost-effective in specific niche cases.

The operation of both MOSFETs and JFETs can be analyzed using various models. Simplified models provide intuitive understanding of basic device behaviour. For detailed circuit simulations, more complex models incorporate parameters that precisely capture the influence of various effects, including channel length modulation, substrate bias, and temperature dependence. Understanding these models is crucial for accurate circuit design and performance prediction. Furthermore, choosing the correct model for a specific application directly impacts the accuracy of your simulations and ultimately the effectiveness of your circuit design.

In summary, FETs are crucial components in modern electronics, offering advantages in terms of power consumption, input impedance, and scalability compared to BJTs. Both MOSFETs and JFETs have distinct operational principles and characteristics, leading to their application in various circuit designs. The judicious selection of an appropriate FET type, considering its advantages and disadvantages, is essential for optimizing electronic circuit performance. Selecting the proper model for analysis and simulation is similarly critical for obtaining accurate results and efficient circuit designs. Each device has its strengths and weaknesses, and understanding these nuances is critical for optimal electronic circuit design.

5.1 MOSFET Operation

The MOSFET, or Metal-Oxide-Semiconductor Field-Effect Transistor, is a cornerstone of modern electronics, its ubiquitous nature a testament to its versatility and efficiency. Unlike bipolar transistors which rely on current injection to control current flow, the MOSFET leverages an electric field to modulate the conductivity of a channel, providing excellent control and low power consumption. This fundamentally different operating mechanism opens up a wide range of applications from microprocessors to power electronics, highlighting its importance in a broad spectrum of technological advancements. At the heart of the MOSFET lies its unique structure. It comprises a silicon substrate, typically p-type, on which lies a thin layer of silicon dioxide (SiO2) acting as an insulator. This oxide layer separates the gate electrode, a metal layer that controls the channel, from the substrate. Source and drain regions, heavily doped n-type silicon in the case of an n-channel MOSFET, are diffused into the substrate, providing the paths for current flow. The ingenious design is precisely what allows for the gate to exert such powerful control over the flow of charge carriers. A positive voltage applied to the gate attracts electrons towards the silicon-oxide interface, forming a conductive n-type channel. When the voltage exceeds a certain threshold, this channel allows current to flow between the source and drain, acting as a switch. Importantly, the gate's control is fundamentally electrical, requiring very little current, providing a remarkable level of energy efficiency.

Understanding the MOSFET's operational characteristics requires a grasp of its different regions of operation. In the cutoff region, the gate voltage is insufficient to induce a channel, leading to minimal current flow. This acts like an open switch. In the linear or triode region, a channel is created but its conductivity changes with gate voltage; acting like a variable resistor, allowing for a finely tuned current. Finally, in the saturation region, a fully formed channel exists with constant conductivity regardless of changes in drain-source voltage, acting as a constant-current source. The ability to shift seamlessly between these modes allows for complex circuit designs and functionalities.

The threshold voltage, a critical parameter, determines the minimum gate voltage required to form a conductive channel. This voltage is strongly influenced by the doping concentration of the substrate, the oxide thickness, and various process parameters. Precision in controlling this value is crucial for consistent device performance and reliable circuit operation. Slight variations in threshold voltage can impact the MOSFET's switching behavior, ultimately affecting the performance of the overall integrated circuit. Furthermore, understanding the relationship between threshold voltage and temperature is essential for ensuring consistent operation over a wide range of environmental conditions. Careful consideration of this parameter is critical during the design and fabrication processes, impacting device characteristics and the reliability of circuits they are integrated into.

Beyond the fundamental operation, MOSFETs are categorized into several types based on their construction and characteristics. Enhancement-mode MOSFETs require a gate voltage to create a channel, whereas depletion-mode MOSFETs have a built-in channel that is modulated by the gate voltage. Moreover, n-channel and p-channel MOSFETs are differentiated by the type of charge carrier, electrons or holes, that constitute the channel current. The choice of MOSFET type hinges on the specific application requirements, and each type exhibits distinct advantages and limitations in terms of speed, power consumption, and noise characteristics. These features influence various circuit designs, particularly in digital and analog applications.

The MOSFET's impact on modern technology is profound and far-reaching. Its role in integrated circuits is undeniable, forming the foundation of microprocessors, memory chips, and countless other electronic devices. Its high input impedance, low power consumption, and ease of fabrication have made it the device of choice in numerous applications, driving innovation and miniaturization across the electronic landscape. Understanding the complexities of MOSFET operation is key to unlocking further advancements in semiconductor technology. The ever-increasing demand for faster, smaller, and more energy-efficient devices will undoubtedly continue to fuel research and innovation in this critical area. The MOSFET is more than just a component; it's a cornerstone of the digital revolution.

5.2 JFET Operation

JFETs, or Junction Field-Effect Transistors, represent a crucial class of field-effect transistors (FETs). Unlike bipolar junction transistors (BJTs) which rely on current injection to control current flow, JFETs utilize an electric field to modulate the conductivity of a channel. This is achieved through a reverse-biased PN junction, thus, understanding the PN junction's behavior is a prerequisite to comprehending JFET operation. Their simplicity and inherent properties make them suitable for various applications, from amplifiers to switches.

The basic structure of an n-channel JFET comprises three terminals: the source (S), drain (D), and gate (G). A highly doped n-type semiconductor forms the channel connecting the source and drain. This channel is flanked by two p-type regions forming the gate. The gate regions are connected, forming a single gate terminal. A p-channel JFET exhibits a similar structure but with the roles of p-type and n-type semiconductor materials reversed.

Operation begins with the application of a voltage between the source and drain, establishing a current flow through the n-type channel. This current, referred to as the drain current (ID), is directly proportional to the voltage across the channel (VDS). However, applying a negative voltage (for an n-channel JFET) to the gate terminal modifies the depletion region between the gate and channel. This depletion region expands into the channel, effectively reducing its conductive cross-sectional area.
As the gate-source voltage (VGS) becomes more negative, the depletion region widens, further constricting the channel and consequently reducing the drain current. At a specific VGS value, known as the pinch-off voltage (VP), the channel is completely closed, effectively turning off the JFET. This characteristic makes JFETs ideal for switching applications where precise current control is needed. Beyond pinch-off, further increasing the negative gate-source voltage results in a nearly constant drain current, showing the saturation region.
The drain current is not entirely unaffected by VDS in saturation, increasing slightly with it. This is due to the phenomenon of channel length modulation. The depletion region extends further at the drain end of the channel for increased VDS, leading to a slight increase in ID. However, this dependence is rather small compared to the impact of VGS.
The JFET's characteristic curve, plotting ID against VDS for different values of VGS, illustrates this behavior. It shows a steep rise in ID at lower VDS values, followed by a flattening in the saturation region. The curves shift downwards with increasingly negative VGS, showing reduced current for the same drain voltage. These curves are essential for understanding JFET behavior and choosing appropriate operating conditions for a circuit.
Mathematical expressions, derived from semiconductor physics principles, model the JFET's current-voltage characteristics with impressive accuracy. These equations encompass the relationships between ID, VGS, and VDS, factoring in the device's physical parameters. While their precise derivation requires detailed knowledge of

semiconductor physics, understanding the interplay between these variables is crucial for circuit design. The Shockley equation for JFET is a very known model used to determine the behavior of the device.

JFETs boast several advantages over BJTs. They exhibit high input impedance, meaning they draw minimal current from the input signal source. This is because the gate is reverse-biased, minimizing the gate current. This high input impedance is especially beneficial in applications where signal integrity is paramount. They also generally exhibit lower noise compared to BJTs, making them ideal for sensitive analog applications. Additionally, they have simpler fabrication processes compared to more complex transistor structures.

However, JFETs also possess certain limitations. Their transconductance, which characterizes the amplification capability, is generally lower than that of BJTs. Furthermore, their range of operation is usually more limited than BJTs, showing less flexibility in the biasing conditions. Therefore, selecting the appropriate device often depends on specific application requirements and the associated trade-offs. A careful consideration of these factors is paramount for successful circuit design and performance. Understanding these subtleties is vital for efficient circuit design and optimization. Precise analysis using appropriate models is essential for accurate predictions and performance optimization.

5.3 Comparison of FET types

Metal-oxide-semiconductor field-effect transistors (MOSFETs) and junction field-effect transistors (JFETs) represent two major classes of field-effect transistors (FETs), each possessing unique characteristics that make them suitable for different applications. Understanding these differences is crucial for effective circuit design. MOSFETs utilize a gate insulator, typically silicon dioxide, to electrically isolate the gate from the channel, leading to high input impedance and low power consumption. This isolation allows for a more robust and predictable control of the channel conductivity. In contrast, JFETs rely on a direct junction between the gate and channel, resulting in a lower input impedance and potentially higher leakage currents. This difference significantly impacts their performance in various circuit configurations.

The operational principles of MOSFETs and JFETs diverge fundamentally. MOSFETs are voltage-controlled devices; the gate voltage modulates the channel conductivity by inducing or depleting charge carriers in the channel region. This modulation results in a change in the drain-source current. This voltage control offers advantages in terms of precise current regulation and ease of integration into complex circuits. JFETs, on the other hand, are current-controlled devices; the gate-source voltage controls the depletion region width, thus regulating the channel width and subsequently altering the drain-source current. This current control mechanism, however, makes them less intuitive to operate and more sensitive to temperature variations, limiting their use in some high-precision applications.

The construction of these transistors also varies. MOSFETs typically incorporate a metal gate, a silicon dioxide insulator, and a semiconductor channel. The gate insulator provides the high input impedance, a key characteristic of MOSFETs. The MOSFET structure allows for a wider range of fabrication techniques, enabling miniaturization and high-density integration into integrated circuits. JFETs, however, have a simpler structure, consisting of a semiconductor channel and two p-n junctions that form the gate regions. This simpler structure can lead to easier fabrication in some cases, but it also limits their scalability compared to MOSFETs. This direct junction between the gate and channel, while leading to a simpler structure, limits their performance in certain high-frequency applications.

Looking at their respective applications highlights their distinct advantages. MOSFETs dominate in digital circuits due to their high input impedance, low power consumption, and ease of integration. Their ability to switch on and off rapidly makes them ideal for constructing logic gates and memory elements. The low leakage currents are particularly advantageous in low-power applications, such as portable electronics. JFETs, conversely, find more use in analog circuits, especially in situations where a relatively constant drain current is needed. Their less sensitive input voltage characteristics compared to bipolar junction transistors (BJTs) can be beneficial in amplifiers where higher stability is desired.

The differences in their characteristics extend to their manufacturing costs and complexities. While both are mature technologies, MOSFETs, due to the complexity of their fabrication process involving multiple steps to create the oxide layer and finely defined channel regions, generally result in higher manufacturing costs at smaller dimensions, though the economies of scale have lessened these differences. JFETs are generally less complex to produce, which can result in lower costs, especially in low-volume manufacturing. This cost advantage becomes relevant when considering less demanding applications where the performance advantage of MOSFETs isn't necessary. However, in modern integrated circuits, the superior scalability of MOSFETs ultimately leads to lower per-device cost in high-volume production.

Finally, consideration of their operating frequencies reveals another key distinction. MOSFETs, particularly those employing advanced fabrication techniques like FinFETs, generally exhibit higher operating frequencies than JFETs. The greater control of the channel and improved switching speed afforded by the insulated gate leads to better performance in high-speed applications. JFETs, with their inherently slower response due to the capacitance of the depletion region, are not as well suited for applications that demand high bandwidth or fast switching speeds. This makes MOSFETs the preferred choice in modern high-speed integrated circuits, such as microprocessors and memory chips. The tradeoffs between these devices highlight the importance of understanding application requirements when choosing between MOSFET and JFET technologies. Carefully considering these comparative analyses allows for an optimal choice for specific circuit designs and ultimately results in a more efficient and higher-performing final product. Their distinct operational principles, construction details, applications, and manufacturing considerations ultimately demonstrate the diverse landscape of field-effect transistor technology.

6 Integrated Circuits

The creation of integrated circuits (ICs), also known as microchips, represents a monumental achievement in miniaturization and efficiency. Think of it: billions of transistors, resistors, and capacitors, all crammed onto a tiny silicon wafer, working together in perfect harmony. This incredible feat is made possible through a series of sophisticated fabrication techniques, transforming theoretical circuit designs into functional, real-world devices. The process begins with the creation of a silicon wafer, a highly purified disc of silicon, incredibly smooth and almost flawless at the atomic level. This is the foundation upon which the entire IC will be built.

Photolithography is the cornerstone of IC fabrication. This process involves coating the silicon wafer with a photosensitive material, called photoresist. A mask, a carefully designed template containing the circuit pattern, is placed over the photoresist. Ultraviolet light then shines through the mask, exposing specific areas of the photoresist. The exposed areas are then chemically developed, removing the exposed photoresist. This creates the pattern of the circuit on the wafer, essentially etching the future components onto the silicon's surface. The precision required is astonishing, measured in nanometers - billions of a meter!

After the photolithography, a series of etching processes refine the patterns created. This involves using chemicals or plasma to remove the exposed silicon, creating the three-dimensional structures of transistors and other components. Multiple layers of photolithography and etching are typically needed to build the complex structures found in modern ICs. Each layer adds another level of complexity, building upon the previous ones. This intricate, multi-step process requires extreme cleanliness and control; even a tiny speck of dust can ruin the entire wafer.

Doping plays a vital role in IC fabrication, modifying the electrical properties of specific silicon regions. This involves introducing impurity atoms, such as boron or phosphorus, into the silicon lattice. Boron introduces "p-type" regions with an excess of holes, while phosphorus creates "n-type" regions with an excess of electrons. These differently doped regions are essential for the creation of transistors, diodes, and other active components in the IC. Careful control of doping concentration is crucial for achieving the desired electrical characteristics. The precision is remarkable—the process creates regions with controlled atomic-scale variation in the composition and therefore properties of the material.

Following the fabrication of individual components, metallization interconnects these components, forming the electrical pathways necessary for the IC to function. This step involves depositing thin layers of metal, typically aluminum or copper, onto the wafer, creating a complex network of wires connecting transistors and other components. The metallization process must ensure that the connections are reliable and have low resistance. Without this crucial step, our billions of transistors would be isolated and unable to operate together as a cohesive circuit.

The completed wafer is then tested rigorously to identify any faulty chips. Each individual die (a single IC) is inspected to ensure it meets specifications before being packaged. Only after passing stringent testing are the chips packaged, protecting them from the environment and providing connections to the outside world. This packaging is as sophisticated as the chip itself, ensuring efficient heat dissipation and robust mechanical protection.

Circuit design plays a crucial role before the actual fabrication begins. Engineers use specialized software to design the complex layout of the IC, placing transistors, resistors, and other components in optimal positions for minimal power consumption and maximum performance. Efficient design considerations also include minimizing signal delays and noise, crucial for high-speed operation. The challenge here lies in not only creating the circuit's functionality, but also ensuring that it's physically realizable on a tiny chip with limited space for wiring.

The design must account for various parameters, including the physical dimensions of the components, the distances between them, and the expected current flows. Complex simulations are used to test the design's performance before it is sent for fabrication. This verification stage allows for iterative design improvements, reducing manufacturing costs and preventing design flaws that might arise only during the fabrication process. Finally, the entire process demonstrates the power of collaboration between physics, chemistry, and engineering - a true testament to human ingenuity. The intricate dance of atoms and electrons, orchestrated with precision, creates a technology that drives our modern world.

6.1 Fabrication Techniques

The creation of integrated circuits (ICs) relies on a complex interplay of techniques, each meticulously designed to pattern and modify the semiconductor material. This process, often referred to as microfabrication, starts with a highly purified silicon wafer, a thin, circular slice of single-crystal silicon, the foundation upon which the intricate circuitry is built. The wafer's surface must be exceptionally smooth and clean, requiring rigorous cleaning processes to remove any contaminants that could hinder the subsequent steps.

Photolithography, arguably the most crucial fabrication technique, involves transferring a circuit pattern from a photomask onto the wafer. The photomask is essentially a transparent plate with opaque regions defining the desired circuit layout. Ultraviolet light shines through the mask, exposing a light-sensitive material, photoresist, coated on the wafer. Areas exposed to the light undergo a chemical change, allowing selective etching of the silicon or deposition of new materials. The resolution of the photolithography process is critical, determining the minimum feature size that can be created, thus directly impacting the density and complexity of the integrated circuit. This intricacy is essential for accommodating billions of transistors within the confines of a single chip, a feat unimaginable without meticulous photolithographic techniques.

Following photolithography comes etching, a process that removes material from the exposed areas of the wafer. Several etching techniques exist, each with its advantages and drawbacks. Wet etching utilizes chemical solutions to dissolve the exposed silicon, offering excellent selectivity but potentially lacking precision in creating intricate patterns. Dry etching, on the other hand, employs plasma or reactive ion beams, offering greater control and higher resolution, enabling the fabrication of finer features essential for modern high-density ICs. The choice of etching technique is dictated by the specific requirements of the circuit design and the materials being processed, a careful selection that impacts the overall performance and yield of the IC production.

Once the desired pattern is etched, it is often necessary to deposit additional layers of material to form the various components of the circuit. This deposition process involves adding layers of materials, such as silicon dioxide, silicon nitride, or metals like aluminum or copper, to create insulating layers, conducting pathways, or gate electrodes. Different deposition techniques exist, including chemical vapor deposition (CVD), physical vapor deposition (PVD), and atomic layer deposition (ALD), each offering unique characteristics in terms of material quality, deposition rate, and conformability to complex three-dimensional structures. The selection of a particular deposition method hinges on the desired properties of the deposited layer and the requirements of the fabrication process, each technique tailored for optimal results.

After deposition, further processing steps might include ion implantation, a crucial technique for precisely controlling the electrical properties of specific regions of the silicon. This process involves accelerating ions of dopant atoms into the wafer, altering the silicon's conductivity to create regions of n-type or p-type material, essential for building transistors and other semiconductor devices. The precise control of ion implantation parameters, such as ion energy and dose, directly impacts the electrical characteristics of the device, making this a critically important step in the fabrication process. Ion implantation allows us to control the semiconductor's conductivity with exceptional precision, shaping its behavior to meet the specific requirements of the circuit.

Finally, the completed IC undergoes testing to ensure its functionality and performance. Numerous tests are performed to verify the integrity of individual components and the overall circuit operation. This meticulous testing process is crucial to identify any defects or failures before the chip is packaged and shipped, ensuring the reliability and performance of the final product. These tests not only validate the functioning of each element but also reveal the effectiveness of the entire fabrication process, offering crucial insights into optimizing future production runs. The meticulous integration of these techniques allows engineers to create highly complex circuits, the backbone of modern electronics, pushing the boundaries of technological advancement. Each step represents a testament to human ingenuity and precision engineering.

6.2 Circuit Design

Designing efficient and effective circuits requires a deep understanding of semiconductor device behavior, which we've explored in previous chapters. Successful circuit design isn't just about connecting components; it's about carefully considering their interactions and leveraging their properties to achieve a specific function. This involves understanding the limitations of individual components and how those limitations can be mitigated. For instance, the operating temperature range of transistors can significantly impact circuit performance, often requiring careful heat dissipation strategies to prevent damage or performance degradation. A poorly designed circuit might exhibit unexpected behaviors at elevated temperatures, leading to errors or malfunctions. Therefore, thermal considerations are paramount in practical circuit design.

Let's consider a simple amplifier circuit. You might use a bipolar junction transistor (BJT) to amplify a weak signal. Choosing the right BJT is crucial; its gain, bandwidth, and power-handling capacity all determine its suitability. You'll also need to select appropriate resistor values to set the operating point (bias) of the transistor, ensuring it operates in its active region for linear amplification. Improper biasing leads to distortion or even complete signal cutoff. A key aspect here is understanding how transistor parameters - such as beta (current gain) - vary with temperature and manufacturing tolerances. Account for these variations through careful component selection and possibly incorporating additional circuitry for temperature compensation or robust bias stabilization. These subtleties distinguish a functional circuit from one that performs reliably in diverse operational conditions.

Next, the coupling capacitors play a critical role. They allow the amplified signal to pass through while blocking any DC bias, which is essential for proper signal transmission. Choosing the wrong capacitor value could severely attenuate or distort the signal. This relates directly to the capacitor's impedance, and the frequency response of the entire circuit hinges on the careful selection of capacitors and resistors to ensure the desired signal gain and frequency response. If the values are not accurately chosen, you will end up with a circuit that is not capable of reaching its intended signal gain without undesirable distortions and attenuation at various frequencies. It's not just about meeting a specification sheet; it's about understanding the underlying physics that governs component behavior.

Moving beyond basic amplification, let's delve into more complex circuits involving integrated circuits (ICs). Designing with ICs involves more than merely connecting pins; it demands a thorough understanding of the IC's data sheet. This includes understanding parameters like input impedance, output impedance, and gain bandwidth product - all of which will directly impact the performance of the overall circuit. Moreover, you must account for noise and interference, and consider techniques like shielding and proper grounding to minimize signal degradation. The placement of components also matters; careful layout design minimizes unwanted capacitive and inductive coupling that can lead to instability or unexpected behavior. The principles of signal integrity are crucial here, especially in high-speed digital circuits.

Finally, consider the testing and debugging process. This isn't just an afterthought; it's an integral part of the design cycle. Thorough testing validates the design's functionality and performance and identifies any unforeseen issues. Simulations are valuable tools for verifying functionality and identifying potential problems before committing to physical prototyping. However, they don't replace empirical testing, which remains essential to fully validate and optimize the circuit's reliability under diverse conditions and real-world influences. This stage relies on the use of appropriate test equipment and methodologies to ensure the circuit meets the desired specifications. Remember, a well-designed circuit should not only function as intended but should also be robust and reliable. This involves not only theoretical understanding, but also practical skills in breadboarding, testing and troubleshooting circuits to enhance their longevity and operational efficacy.

7 Advanced Topics

High-frequency operation presents unique challenges to semiconductor devices. At gigahertz frequencies and beyond, the transit time of charge carriers across the device becomes comparable to or even exceeds the period of the signal. This leads to a reduction in gain and increased noise, effects not significant at lower frequencies. Careful consideration of parasitic capacitances and inductances, inherent to the device structure and packaging, is critical. Minimizing these parasitic elements through optimized designs and advanced fabrication techniques is crucial for maintaining performance at high frequencies. Understanding the frequency response, often characterized by parameters such as cutoff frequency and bandwidth, is essential for selecting appropriate devices and designing high-speed circuits. Advanced modeling techniques, often employing electromagnetic simulations, are frequently utilized to predict and optimize device behavior at high frequencies. This necessitates a comprehensive understanding of both the semiconductor physics and electromagnetic theory. Furthermore, the development of novel materials and device architectures specifically engineered for high-frequency applications continues to be an active area of research, pushing the boundaries of what's possible in high-speed electronics.

Heterojunctions, where two different semiconductor materials are joined, offer significant advantages over homojunctions (junctions between the same material). The most notable is the ability to engineer the band gap and band offsets at the interface. This precise control allows for the creation of devices with enhanced performance characteristics. For example, a heterojunction can create a built-in potential barrier that effectively confines charge carriers, improving device efficiency. This principle finds application in high-electron-mobility transistors (HEMTs), exhibiting extremely high electron mobility compared to conventional FETs. The bandgap engineering also allows for the creation of devices with tailored optical properties, finding extensive use in optoelectronic devices such as lasers and light-emitting diodes (LEDs). Furthermore, heterojunctions can be used to reduce leakage currents, which is particularly beneficial in high-power or high-temperature applications. The design and fabrication of heterojunctions require sophisticated control of material growth techniques such as molecular beam epitaxy (MBE) and metalorganic chemical vapor deposition (MOCVD) to achieve precisely controlled interfaces. These techniques allow the formation of highly abrupt and atomically sharp interfaces, essential for optimizing device performance.

Nanotechnology has revolutionized the semiconductor industry, enabling the creation of devices with unprecedented miniaturization and functionality. The manipulation of materials at the nanoscale unlocks new possibilities, including the fabrication of nanowires, nanotubes, and quantum dots. Nanowires, for example, exhibit unique electrical and optical properties due to their high surface-to-volume ratio and quantum confinement effects. Similarly, carbon nanotubes, known for their exceptional mechanical strength and electrical conductivity, offer promise in creating ultra-high-frequency transistors and interconnects. Quantum dots, which confine electrons in all three spatial dimensions, exhibit size-dependent energy levels, finding applications in quantum computing and single-photon sources. Challenges associated with nanotechnology include precise control of fabrication processes, ensuring uniform nanostructure properties, and integration of nanomaterials into existing semiconductor fabrication processes. However, the potential benefits are significant, driving intense research and development efforts towards exploiting the unique properties of nanomaterials in advanced semiconductor devices and circuits.

Semiconductors play a crucial role in the creation of optical devices, forming the foundation for lasers, photodetectors, and modulators. Semiconductor lasers, in particular, leverage the stimulated emission of light from a population inversion within a semiconductor material. The precise control of carrier concentration and energy band structure allows for the generation of light with specific wavelengths and power levels, impacting applications from optical communications to laser surgery. Photodetectors, on the other hand, utilize the photoelectric effect to convert light into electrical signals, essential components in optical communication systems and imaging sensors. Semiconductor modulators enable the manipulation of light intensity or phase, enabling crucial functionalities in optical networks and data transmission. The development of advanced materials and device architectures continually enhances the performance of these optical components, enabling higher speeds, improved sensitivity, and greater energy efficiency. The integration of these semiconductor-based optical devices with existing electronic circuits is a key challenge, leading to significant research efforts in developing hybrid photonic-electronic integrated circuits (OEICs).

The future of semiconductor technology promises exciting advancements. Research in new materials, such as two-dimensional materials like graphene and transition metal dichalcogenides, holds the potential for creating devices with superior performance characteristics compared to silicon-based technology. Further miniaturization, pushing beyond the limits of conventional lithographic techniques, will require the exploration of novel fabrication methods such as self-assembly and nanoimprint lithography. Quantum computing, leveraging the principles of quantum mechanics, is poised to revolutionize computational capabilities, with semiconductors playing a central role in the development of quantum bits (qubits). Moreover, integrating artificial intelligence into the design and optimization of semiconductor devices promises to accelerate innovation and improve device performance. The development of energy-efficient and environmentally friendly semiconductor manufacturing processes is another critical area of research. As we continue to push the boundaries of what's possible in semiconductor technology, the potential applications and societal impact are immense, shaping the future of electronics and beyond.

7.1 High-Frequency Effects

At high frequencies, the seemingly simple behavior of semiconductor devices we've examined so far becomes significantly more complex. The transit time of charge carriers across device junctions, previously negligible, becomes a critical factor influencing performance. This leads to a reduction in gain and an increase in noise, significantly impacting circuit functionality at gigahertz frequencies and beyond. Understanding these high-frequency limitations is vital for designing high-speed circuits and systems.

Consider, for instance, a simple bipolar junction transistor (BJT). At low frequencies, the current gain is relatively constant. However, as the frequency increases, the carriers injected into the base region might not have enough time to reach the collector before the input signal changes, significantly decreasing the effectiveness of current amplification. This effect, known as the base transit time effect, ultimately limits the high-frequency response of the BJT. The inherent capacitance present within the device also plays a significant role. Capacitance is primarily between the base and collector, and between the base and emitter. These parasitic capacitances, often ignored at lower frequencies, form low-pass filters, attenuating higher-frequency components of the signal. A thorough understanding of the frequency dependence of these capacitances is essential for accurate high-frequency circuit modeling.

The situation is similar, though not identical, in field-effect transistors (FETs). In MOSFETs, for example, the gate-to-source and gate-to-drain capacitances become dominant at high frequencies. These capacitances, along with the inherent resistance of the channel, determine the device's high-frequency response. The parasitic capacitances, as in BJTs, again act as low-pass filters, causing signal attenuation above a certain frequency. Moreover, the channel transit time - the time it takes for electrons to traverse the channel from source to drain - becomes a crucial factor, limiting the speed at which the device can respond to changes in the input signal. This high-frequency behavior dictates the maximum operating frequency of the device, thereby limiting its use in certain applications. This is particularly relevant when designing high-speed digital circuits where signal transitions need to occur as rapidly as possible.

Another critical factor to consider is the effect of high frequencies on noise. At higher frequencies, various noise sources within the semiconductor device become amplified, leading to a significant reduction in the signal-to-noise ratio. Thermal noise, shot noise, and flicker noise, previously manageable at low frequencies, increase considerably with frequency. The parasitic capacitances further enhance the effect of thermal noise. A comprehensive understanding of these noise sources and their frequency dependence is essential for optimizing circuit design at high frequencies, thereby ensuring an acceptable signal integrity in the finished product. Advanced circuit design techniques are employed to mitigate these issues, such as using special configurations and compensation networks to minimize unwanted signal attenuation and noise amplification. Careful selection of the semiconductor materials, geometries, and fabrication processes are also crucial for improving high-frequency performance.

Furthermore, interconnections within integrated circuits themselves become significant transmission lines at high frequencies. The inductance and capacitance of these interconnections cannot be ignored, and the characteristic impedance of the lines, the velocity of propagation of signals, and the reflections from mismatches have to be carefully considered. These parasitic effects are particularly impactful in high-speed digital and microwave circuits. Proper consideration of these interconnect effects, including careful layout design and the use of advanced signal routing strategies, becomes critical to ensure signal integrity and minimize signal degradation at high frequencies. The accurate modeling of these effects requires specialized software and sophisticated electromagnetic simulations. Hence, mastering these techniques is essential for today's designers of high-speed circuits and systems. Moreover, these effects are only magnified as circuit feature sizes decrease, a direct consequence of the miniaturization of semiconductor technology.

The consequences of neglecting these high-frequency effects are significant. Incorrectly predicting the behavior of a device at high frequencies can lead to unexpected performance limitations, signal degradation, and ultimately, circuit malfunction. Detailed high-frequency modeling, using advanced simulation tools and a thorough understanding of device physics, is crucial for successful high-speed circuit design. This involves a deep understanding of not only the device physics but also advanced electromagnetic simulations, particularly when considering the intricate layout and interconnect effects that become critically important at high frequencies. Thus, the study of high-frequency effects extends beyond the individual transistor and encompasses the complete integrated circuit and its interaction with its environment.

7.2 Heterojunctions

Heterojunctions represent a significant advancement in semiconductor technology, offering capabilities beyond those achievable with homojunctions (junctions between the same semiconductor material). Imagine a symphony orchestra, where each instrument plays its unique part - a heterojunction is like combining instruments with vastly different tonal qualities to create a richer, more complex sound. This enhanced functionality stems from the ability to tailor the bandgap and other material properties by combining different semiconductors.

The key advantage of a heterojunction lies in its ability to engineer the energy band alignment at the interface between two dissimilar materials. This precise control allows for the creation of devices with superior performance characteristics. For example, consider a heterojunction laser diode. By cleverly selecting materials with different bandgaps, we can create a structure where electrons and holes are efficiently confined to a specific region, leading to a significant increase in the radiative recombination rate and therefore brighter, more efficient light emission. This precise control over energy levels isn't possible with homojunctions.

The design and fabrication of heterojunctions require a deep understanding of material science and semiconductor physics. You need to carefully choose materials with compatible lattice constants to minimize strain and defects at the interface. This is critical because defects can act as non-radiative recombination centers, reducing the efficiency of the device. The selection also depends on the specific application. For high-frequency applications, materials with high electron mobility are preferred. For optoelectronic devices, the bandgap of the materials must be chosen to match the desired wavelength of light emission or absorption.

One of the most important applications of heterojunctions is in high-electron-mobility transistors (HEMTs). In a HEMT, a heterojunction is used to create a two-dimensional electron gas (2DEG) at the interface. The 2DEG has extremely high electron mobility due to reduced scattering from impurities and lattice vibrations. This results in HEMTs exhibiting superior high-frequency performance compared to conventional transistors, making them crucial components in high-speed communication systems and other applications demanding speed. This is a testament to the ingenuity of harnessing the unique properties of material interfaces.

Another striking example is the use of heterojunctions in quantum well lasers. These lasers utilize multiple quantum wells, which are thin layers of one semiconductor material sandwiched between layers of another material with a larger bandgap. The quantum wells confine electrons and holes, leading to a dramatic increase in the radiative recombination rate and reduced threshold current. This translates into smaller, more energy-efficient lasers, which find widespread use in optical communication and data storage technologies. The precision of quantum confinement within these structures

truly showcases the power of heterojunction design.
Beyond these applications, heterojunctions are vital in various other semiconductor devices, including solar cells, high-power transistors, and light-emitting diodes (LEDs). In solar cells, heterojunctions can improve the efficiency of light absorption and charge separation. In high-power transistors, they enhance the breakdown voltage and current-carrying capacity. Similarly, heterojunction LEDs exhibit better efficiency and color purity. This versatility underscores the importance of heterojunctions as a foundational element of modern semiconductor technology. Their impact is far-reaching and crucial.
The study of heterojunctions is an active field of research, with ongoing efforts to develop new material combinations and fabrication techniques. Researchers are exploring novel materials and architectures to further improve the performance of heterojunction devices. The pursuit of higher efficiencies, faster speeds, and smaller sizes is a continuous quest. For instance, the integration of nanomaterials with heterojunctions promises to unlock even more sophisticated functionalities, opening exciting possibilities for future applications.
In conclusion, heterojunctions offer a powerful tool for manipulating the electronic and optical properties of semiconductors. The ability to precisely control the energy band alignment at the interface between different materials allows for the creation of devices with significantly enhanced performance characteristics. Understanding the principles of heterojunction design and fabrication is crucial for anyone seeking to delve deeper into the realm of advanced semiconductor devices. The future of semiconductor technology is intrinsically linked to the continued exploration and innovation in heterojunction research. The possibilities are vast and the potential for groundbreaking discoveries remains substantial.

7.3 Nanotechnology

The manipulation of matter at the atomic and molecular level—that's the essence of nanotechnology. Its relevance to semiconductor physics is profound, offering pathways to create devices with unprecedented capabilities. Imagine transistors smaller than a virus, circuits woven from single molecules, and computers with processing speeds exceeding anything we currently possess. These are not just futuristic dreams; they are the tangible goals driven by the principles of nanotechnology. Understanding its core concepts is key to appreciating its transformative potential in semiconductor technology.

One crucial application lies in the creation of quantum dots. These are tiny semiconductor crystals, typically a few nanometers in diameter, that exhibit quantum mechanical properties. Their size-dependent band gap allows for precise control over their optical and electronic characteristics, making them ideal for various applications. From highly efficient LEDs to advanced lasers and solar cells, the potential for improved performance is immense. Their use in advanced display technologies already illustrates the exciting possibilities. Researchers continue exploring their potential in quantum computing, leveraging their unique quantum properties for computation.

Beyond quantum dots, nanowires present another exciting frontier. These one-dimensional structures offer superior control over electron transport compared to bulk materials. Their small size reduces scattering effects, resulting in higher electron mobility. Consequently, they are incredibly promising for creating faster and more energy-efficient transistors and other electronic components. Imagine creating high-performance electronics with reduced power consumption—this is the promise of nanowires. This is important because it directly addresses the limitations of traditional CMOS technology and points toward the evolution of electronics toward even more compact and effective architectures. Their potential in sensing applications also remains largely unexplored, holding further advancements for the field.

Carbon nanotubes, cylindrical structures of carbon atoms, are another remarkable material in nanotechnology. Their exceptional mechanical strength, high electrical conductivity, and unique electronic properties make them ideal for a variety of applications. They could potentially replace traditional silicon in transistors, leading to faster switching speeds and higher current carrying capacities. Their lightweight and flexible nature also opens up new possibilities in flexible electronics. This flexibility is especially relevant for wearable and implantable devices, transforming how we interact with technology and monitor health.

However, the integration of nanomaterials into semiconductor devices presents significant challenges. Precise control over the synthesis and placement of nanostructures is crucial. Developing scalable and cost-effective manufacturing techniques that maintain the high quality and consistency of these structures remains a major hurdle. Understanding and managing the interactions between nanomaterials and their surrounding environment is equally critical. This aspect requires advancements in material science and precise characterization techniques to ensure reliable and long-lasting devices.

The future of nanotechnology in semiconductors lies in overcoming these challenges. Ongoing research focuses on developing new fabrication techniques, such as self-assembly and nanolithography, to create more complex and sophisticated nanostructures. Further progress in characterization techniques is crucial for understanding the fundamental properties of these materials and optimizing their performance in devices. A deeper understanding of nanoscale physics will facilitate the design of novel devices with unprecedented capabilities and unlock the vast potential of nanotechnology within the semiconductor industry. The synergy between these diverse areas promises to reshape the semiconductor landscape for generations to come. The prospect of fully realizing this potential will inevitably reshape how we design, manufacture and utilize semiconductor components and systems in our everyday lives. The implications for energy efficiency, computational power, and data storage are simply profound, demanding further investigation and technological advancement. The ongoing development of improved fabrication methods and characterization tools continues to fuel the exploration of nanotechnology's promise in the semiconductor world.

7.4 Optical Devices

The interaction of light with semiconductors forms the basis for a vast array of optical devices. These devices, unlike their purely electronic counterparts, leverage the ability of semiconductors to absorb, emit, and modulate light. This characteristic stems from the quantized nature of energy levels within the semiconductor material, allowing for precise control over the interaction with photons. Understanding this interaction is crucial for designing and utilizing these powerful tools.

Semiconductor lasers are prime examples, generating coherent light through stimulated emission. In these devices, electrons are injected into the conduction band, where they recombine with holes in the valence band, releasing photons. The emitted photons then stimulate further recombination events, leading to an avalanche of photons of the same wavelength and phase—coherent light. The laser's wavelength is determined by the semiconductor's bandgap, offering precise tunability by selecting appropriate materials or altering doping levels. Laser diodes, a common type of semiconductor laser, are found in many applications, from barcode scanners to fiber optic communication systems, their versatility a testament to the fundamental principles at play.

Photodetectors, on the other hand, convert light into electrical signals. These devices operate on the principle of photoelectric effect, where incident photons excite electrons from the valence band to the conduction band, increasing the material's conductivity. The resulting photocurrent is proportional to the intensity of incident light, enabling the measurement of light levels with high precision. Different semiconductor materials, with their varying bandgaps, are chosen based on the spectral range of interest. Silicon photodiodes, for example, are commonly used in visible-light detection, while other materials are employed in applications requiring sensitivity to infrared or ultraviolet light. These components are indispensable in optical communication, imaging systems, and a myriad of other technologies.

Beyond lasers and photodetectors, semiconductor materials contribute significantly to the development of optical modulators. These devices alter the characteristics of light, such as intensity or phase, in response to an electrical signal. This ability to control light transmission is crucial in high-speed optical communication systems where modulating the light signal allows for efficient data transmission. Electro-optic modulators, relying on the change in the refractive index of a semiconductor under an applied electric field, exemplify this function. The speed and efficiency of these modulators are vital to keep pace with the ever-growing demand for higher bandwidth in communication networks. Careful selection of materials and device design are key to optimizing their performance.

The fabrication of these optical devices often involves techniques similar to those used in integrated circuits, allowing for compact and efficient designs. Photolithography, etching, and doping play crucial roles in creating the intricate structures needed for precise light manipulation. In many cases, these optical components are integrated with electronic circuits on a single chip, leading to the development of optoelectronic integrated circuits (OEICs). This integration simplifies designs, reduces costs, and significantly enhances performance compared to discrete components. This allows for advanced functionalities and miniaturization to an unprecedented level.

Finally, the continuing research and development of new semiconductor materials and device structures are constantly expanding the capabilities of optical devices. Materials with exceptional properties, such as wide bandgaps for high-power applications or specific bandgap engineering for specific wavelengths, are continuously being explored. The exploration of nanostructures and quantum phenomena in semiconductors opens up additional possibilities for device miniaturization and enhanced performance. This relentless pursuit of innovation promises to lead to future generations of even more powerful and efficient optical components, impacting diverse fields such as communication, sensing, and medical imaging. The study of these devices therefore remains a vibrant and crucial area of research with far-reaching implications for technology.

7.5 Future Directions

The relentless pursuit of miniaturization continues to drive innovation in semiconductor technology. Moore's Law, while slowing, remains a powerful guiding principle, pushing us towards ever-smaller, more powerful devices. This necessitates exploration into novel materials and fabrication techniques, transcending the limitations of silicon-based technologies. The integration of new materials, such as gallium nitride (GaN) and silicon carbide (SiC), is already transforming high-power and high-frequency applications, offering superior performance characteristics compared to traditional silicon. This shift is not merely incremental; it represents a paradigm change, opening up new possibilities in areas such as electric vehicles, renewable energy systems, and 5G and beyond communication networks.

Exploring three-dimensional architectures presents another exciting avenue. Traditional planar fabrication methods are reaching their physical limits. Moving beyond two dimensions allows for increased transistor density and reduced interconnect lengths, leading to faster and more energy-efficient chips. This transition requires sophisticated 3D stacking techniques and innovative interconnect solutions, but the potential rewards are immense. Furthermore, the exploration of novel device architectures, beyond the traditional MOSFET, is critical for maintaining the pace of progress. Tunnel FETs (TFETs) and negative capacitance FETs (NCFETs) are promising candidates, offering the potential for lower power consumption and higher performance. However, significant challenges remain in their fabrication and integration into mainstream technologies.

Beyond materials and architecture, advances in computational design and modeling play a crucial role in guiding future developments. Sophisticated simulations allow researchers to explore a vast design space, identify optimal configurations, and accelerate the development cycle. Machine learning techniques can further enhance this process, predicting material properties, optimizing device designs, and even assisting in the automation of fabrication processes. The interplay between theoretical understanding, computational tools, and experimental validation is becoming increasingly important for rapid advancements in the field.

The convergence of semiconductor technology with other disciplines is opening up previously unforeseen opportunities. The integration of photonics with electronics, leading to optoelectronic integrated circuits (OEICs), promises a revolution in data transmission and processing. Combining the advantages of both light and electronic signals allows for faster data transfer speeds, reduced power consumption, and improved signal integrity. This interdisciplinary approach highlights the collaborative nature of innovation in semiconductor technologies. This synergistic approach extends beyond photonics; exploring the convergence with areas like quantum computing promises a truly transformative impact on various aspects of technology.

Quantum computing, while still in its nascent stages, holds the potential to solve problems currently intractable for even the most powerful classical computers. Semiconductors play a vital role in building quantum bits (qubits), the fundamental building blocks of quantum computers. Developing scalable and reliable qubit fabrication techniques using semiconductor materials is crucial for the development of practical quantum computing systems. This is not simply a matter of applying existing knowledge to a new problem; it is an area where fundamental physics and engineering challenges merge to create a whole new field. This integration necessitates a deeper understanding of quantum phenomena, demanding both expertise in quantum mechanics and intricate knowledge of semiconductor physics. Moreover, advancements in materials science are necessary to identify materials better suited to the unique demands of quantum computation.

The future of semiconductor technology is not just about faster and smaller devices; it's about creating devices with enhanced capabilities and functionalities. This requires overcoming substantial hurdles in materials science, device physics, and fabrication techniques. The ongoing research and development in these areas will define the technological landscape for decades to come, influencing not only our electronic devices but also various other critical sectors like healthcare, transportation, and energy. The next generation of semiconductor engineers must embrace interdisciplinary approaches and have a solid grasp of both fundamental and applied knowledge. The combination of theoretical advancements and practical innovation will be vital for propelling the field forward and harnessing the full potential of semiconductor symphonies for years to come. The integration of diverse expertise and the willingness to explore beyond the established paradigms are key ingredients for success in this rapidly evolving field. The future of semiconductor technology is not simply a continuation of existing trends; it is the continuous and unrelenting pursuit of innovation, demanding both visionary thinking and a mastery of fundamental principles.

www.ingramcontent.com/pod-product-compliance
Lightning Source LLC
LaVergne TN
LVHW080531210926
848481LV00021B/2552
9798348504090